Series on Rock and Soil Mechanics

Bins and Bunkers for Handling Bulk Materials

– Practical Design and Techniques –

by

Dr.-Ing. W. Reisner
Dipl.-Ing. M. v. Eisenhart Rothe

Volume 1 (1971) No. 1
January 1971

TRANS TECH PUBLICATIONS

P. O. Box 2981
Cleveland, Ohio 44116
USA

D-3392 Clausthal-Zellerfeld
Adolf-Ey-Str. 1d
Germany

Series on Rock and Soil Mechanics

TRANS TECH PUBLICATIONS

P. O. Box 2981
Cleveland, Ohio 44116
USA

D-3392 Clausthal-Zellerfeld
Adolf-Ey-Str. 1d
Germany

Introduction

Storing and handling of bulk materials have become increasingly important to the mining, mineral processing, chemical, shipping and farming industries over the past years. This is primarily due to the ever increasing capacities of individual plants and installations. Because the prime interest of plant operators was more concentrated on production and capacity, the areas of handling and storing of bulk materials did not receive adequate attention. As bins and bunkers represent only a very small part within a complete industrial complex, the question of their optimal design could only be attacked after a certain rationalisation and automation of the plant was achieved.

The importance of this problem has lead to a number of research works and experimental results. Due to the different industries involved, the various operational conditions and bulk materials, and the different approaches taken by the researchers, the results achieved sofar vary greatly. Nevertheless, they have contributed to a better understanding and lead to some generally applicable knowledge. Unfortunately it was not yet possible to define a concept of ideal bulk masses as in the case of gases or liquids, and it is doubtful if a generally valid theory for bulk materials exists at all.

On the other hand, technically satisfactory results have been found for specific areas. For example the tests on static and dynamic bin loads have lead to some standards and design criteria. The first steps have been taken towards classification of the practically unlimited abundance of bulk materials according to a number of factors. Jenike has developed a method for dimensioning bunkers for cohesive and fine grained materials. A number of theories have been proposed on the mechanical interlocking of coarse materials. However, a more intensive dimensions analysis should be the goal of the future, in order to give the engineer a set of concrete data and formulas for different conditions and materials.

The intensive research during the past five years in the field of materials handling has lead to an intensified exchange of information. During this period alone eight international symposia concerning these problems have been held, and new institutions dealing with material handling in bins and bunkers have been set up in several countries.

The present book has been written in an attempt to dissiminate the widely scattered literature and give the planning engineer a survey on the state of the art and the means to quickly locate articles which deal with his special problems in more detail. It should be kept in mind, though, that this book represents the first attempt in this direction, and that it cannot be expected to be complete in every respect. Comments from scientists, engineers and operators will be appreciated in order to improve a second edition in the future.

The editor and the authors would like to express their thanks and gratitude to the many engineers and companies in this field, who have contributed to this book. Special thanks and sincere appreciation go to Mr. Hendrik Colijn, Materials Handling Consultant, who has acted as co-author of the chapter "Feeding and Recovery" and who has made valuable comments on other sections of the manuscript. The help of Dr. Jenike and Dr. Johanson, Consulting Engineers, who have read and commented on the chapter "Bunker Design Based on the Flow Properties of Bulk Materials" is also gratefully acknowledged.

Editor and Authors

Contents

Page

1. Flow Behaviour of Bulk Solids	1
1.1. Phenomena of gravity flow	1
1.2. Discharge rate	7
1.3. Classification	10
2. Design Based on Flow Properties of Bulk Solids	16
2.1. Design of mass flow bins	17
2.1.1. The shear cell test	18
2.1.2. Determination of necessary data of bulk materials	20
2.1.3. Critical hopper dimensions	24
2.1.4. Range of application	27
2.2. Design of core flow bins	29
2.2.1. Powdered materials	29
2.2.2. Coarse materials	33
3. Loads of Bins and Bunkers	42
3.1. Standards for load calculation	44
3.2. Limitations of the standards	49
3.3. Pressures in mass flow bins	52
4. Shapes of Bins and Bunkers	56
4.1. Capacity of storage	58
4.2. Determination of volume	62
4.3. Cross-section of bins	64
4.4. Shape of outlets and hoppers	65
4.4.1. Outlets in flat-bottom bins	65
4.4.2. Hoppers	68
4.4.3. Inserts	81

	Page
5. Activation of Bins and Bunkers	92
5.1. Liners	93
5.2. Flow inducers	96
5.2.1. Pneumatics	96
5.2.2. Vibrators	100
5.2.2.1. External vibrators	101
5.2.2.2. Vibrating plates	110
5.2.2.3. Suspended vibrating elements	112
5.2.2.4. Vibrating hoppers	113
5.2.3. Mechanical dislodgers	117
6. Feeding and Recovery	119
6.1. Mechanical discharge arrangements	119
6.1.1. Feeder loads	120
6.1.2. Belt feeders	126
6.1.3. Apron feeders	133
6.1.3.1. Bar flight feeders	138
6.1.3.2. Rotary drum, vane and valve feeders	139
6.1.4. Rotary table feeders	141
6.1.4.1. Circular bin dischargers	144
6.1.4.2. Rotary plow feeders	147
6.1.5. Screw feeders	154
6.1.6. Reciprocating plate feeders	160
6.1.7. Vibratory feeders	161
6.1.7.1. Design types	164
6.1.7.2. Supports of trough feeders	169
6.1.7.3. Feeder flow patterns	171
6.1.8. Hopper gates	177
6.1.9. Conclusion	178
6.2. Aeration procedures	185

	Page
7. Functions of Bins and Bunkers	192
7.1. Homogenization of stored materials	193
7.2. Dewatering of stored materials	203
7.3. Bunker conveyors	205
8. Treatment of Bulk Materials	207
8.1. Segregation	208
8.1.1. Cause of segregation	209
8.1.2. Degree of segregation	212
8.1.3. Measures to reduce the degree of segregation	215
8.2. Degradation	220
8.3. Abrasion	226
9. Control and Automatic Equipment	231
9.1. Level measurement	231
9.1.1. Determination of material weight in the bin	234
9.1.2. Electromechanical sounding probes	234
9.1.3. Rotating paddle	234
9.1.4. Diaphragm-type level control	235
9.1.5. Electrical-resistance sensor	235
9.1.6. Level control by conductivity	236
9.1.7. Level control by capacitance	236
9.1.8. Level control with radioactive isotopes	237
9.1.9. Level control by ultrasonics	238
9.2. Flow measurement	240
9.2.1. Volumetric feeders	240
9.2.2. Gravimetric feeders	241
9.2.3. Special methods	244

	Page
9.2.3.1. Optical method	245
9.2.3.2. Nuclear method	246
9.2.3.3. Rotating impeller	246
9.2.3.4. Spiral vane flowmeter	246
9.2.3.5. Inclined plate method	246
9.2.3.6. Method of heat labeling	247
References	249
List of Manufacturers	277

1. Flow Behavior of Bulk Solids

1.1. Phenomena of Gravity Flow

The flow behavior of granular and powdered materials in hoppers has been examined by a number of authors and they have described the flow mechanics in qualitative and quantitative terms. Schwedes (267) has listed these publications in chronological order.

The most frequently applied method to identify the flowzones was to fill up the bunkers with layers of different colors (Fig. 1). The first conclusion of

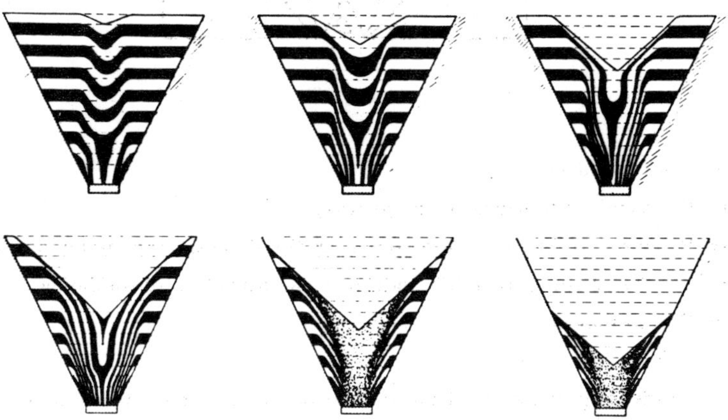

Fig. 1

these experiments is the differentiation between free flowing and non-free flowing bulk materials. Non-free flowing materials are characterized by the phenomena, that the internal forces in particle bed, caused by cohesion, humidity or electrostatic forces, are prevalent over gravity forces. They are capable of developing randomly oriented planes after the flow begins. This leads to the known phenomenon of piping and arching.

These materials are to be handled in suitably calculated hopper constructions or with additional flow-aid devices.

For free flowing materials, Brown and Hawksley (56) described the following flow system near the outlet.

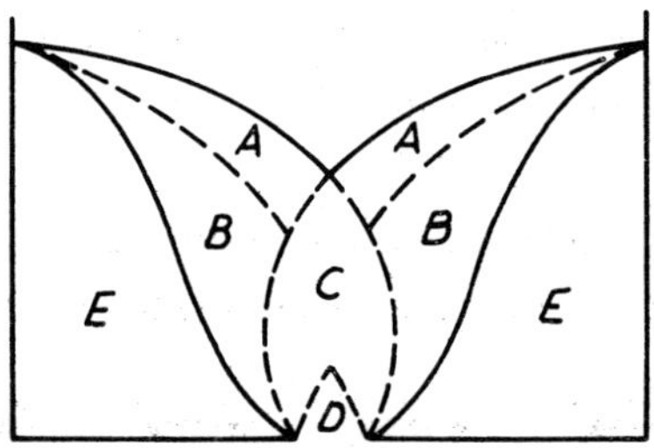

Fig. 2

Zone D = free fall,

Zone C = vertical motions of the grains,

Zone B = slow sliding over the zone E in the direction of the hopper center,

Zone A = fast sliding over the zone B in the direction of the hopper center,

Zone E = no motion.

Further investigations about the formation of flow zones were carried out by Kvapil (163), (166), (167), (170).

Fig. 3 shows two zones and different kinds of motions.

E_N = zone in which primary (vertical) and secondary (rotational) motion takes place,

E_G = socalled boundary-ellipsoid zone beyond which no motion takes place.

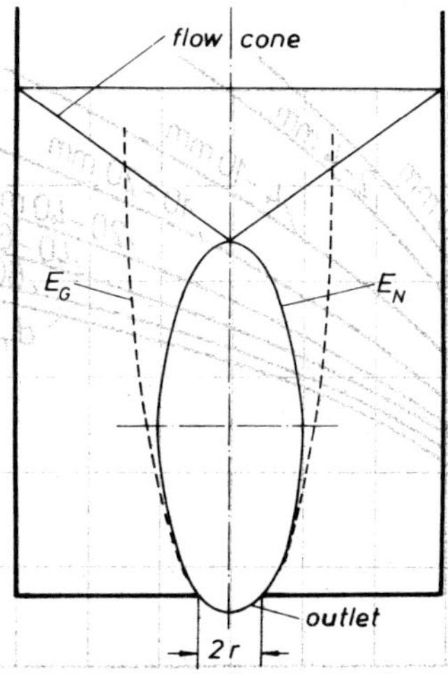

Fig. 3

Both zones continue to develop in a practically constant ratio of 1 : 15, till E_N has reached the surface. Kvapil apprehended that the formation of these zones were dependent upon the material properties, and tried to demonstrate this dependence experimentally. In an example shown in Fig. 4 (169), the height of the ellipsoid E_N is plotted versus its volume for various grain-sizes. The flow pattern confined by zones is usually defined in the literature as funnel or core flow. The emptying of the whole bunker is possible, if the hopper inclination is selected to be greater than the angle of wall friction. The experiments by Kvapil have further lead to some very clear conceptions about the mass flow conditions. Mass flow takes place within the boundary of the ellipsoid E_G. When a first-in/first-out flow pattern is required, one must contour the shape of the bunker to that of the ellipsoid i. e. construct relatively tall hoppers with suitable inclination of the bottom. The design method for this was developed by

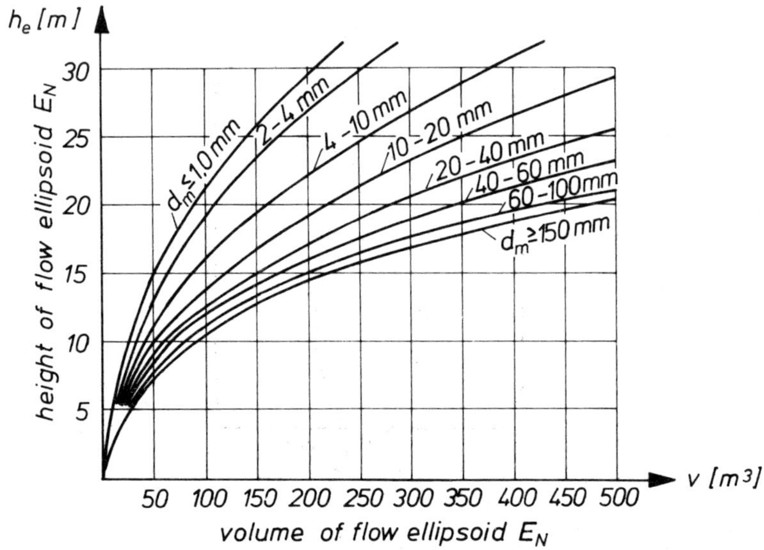

Fig. 4

Jenike (132), (133), also see Chapter 2.1.

With help of X-ray radiographs Athey, Cutress and Pulfer (23), (79) have been able to make the phenomena within the flowing zones visible. After the stationary state of flow is reached, i. e. after the surface is in motion, discontinuous lines of lower powder density are formed which move towards the outlet.

The flow takes place as a continuous shearing along certain slip lines and/or constant building up and collapsing of unstable arches. This was mainly valid for the powder tested. According to Cutress and Pulfer the X-ray technique can be extended to investigate the formation of arches and pipes and the boundaries of flow in non-mass flow systems. The Russian authors in particular, for example Keneman (149), (150), indicate a dynamic arch above the opening in case of free flowing coarse materials. The flow occurs above the arch in coherent layers and below it in free fall. However, recent investigations (37), (40) on flowing pebble beds have shown, that a change occurs in the bulk density in the cylindrical bunker part itself, so that the original state of equilibrium would be changed. These measurements were performed with sensitive probes. When no discontinuity of density could be observed with this method, the density distribution in the pebble bed, in form of arching equi-density curves shows a markable similarity with the results of Cutress and Pulfer (79).

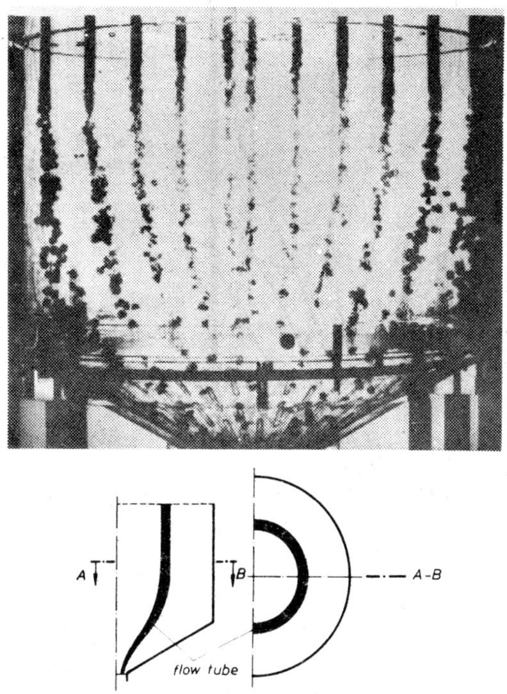

Fig. 5

Very free flowing solids, for instance pebble beds, have a flow pattern which cannot be defined exactly either as mass flow or core flow. Nearly the whole bin volume is activated i. e. motion takes place also at the walls (Fig. 5) (39). This is the definition for mass flow by Jenike. On the other hand the particle velocities in the area between the bin axis and the wall vary in a range of 1 to 20 (see the velocity distribution in Fig. 6). That means that simultaneously a core flow takes place, as there is a faster

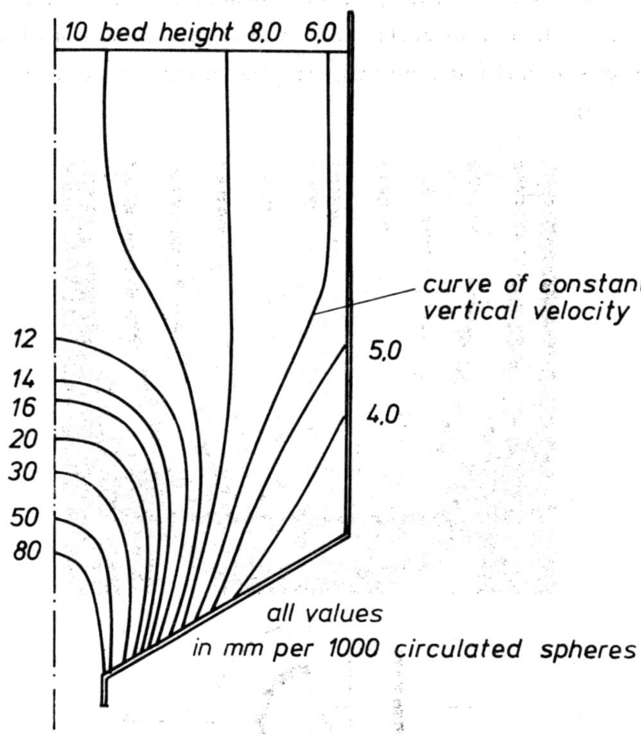

Fig. 6

flow in the center than on the wall. Based on these tests Bedenig (39) defined the flow-tube model to characterize the flow behavior of these materials. If the path curves of the particles are known (see Fig. 5) it is possible to calculate the velocity distribution within the bin by means of continuity condition. On the basis of this flow-tube model Bedenig has drawn up a computer program. Similar tests and calculations to determine the velocity or residence time distribution within a bin were performed by Novosad (212), (213) and Pariseau (220), (221).

1.2. Discharge Rate

A very important and also critical factor for handling of bulk materials in bunkers is the discharge rate. The only effective and well defined possibility to regulate the quantity of efflux is a suitable dimension of the outlet cross-section. The relationship can be formulated as (267)

$$\dot{M} \approx d^{2.5}$$

But the possibility of allowing the material to discharge freely with the required discharge rate is extremely limited

a) due to the critical outlet diameter determined by grain composition and shearing strength of the material (Chapters 2.1.3. and 2.2.2.),
b) due to design considerations, as too high discharge rates cannot be handled anymore by simple locking devices,
c) because of operating reasons e. g. when a variation of the discharge rate or an exact flow control is required.

In these cases at least a certain minimum outlet area must be guaranteed, whereas the actual construction depends primarily upon the requirements of the discharge assembly. In order to achieve optimum flow conditions, the outlet has to be selected as large as possible, but in any case the critical diameter must be maintained to guarantee continuous flow. The free flow discharge thus remains limited to individual cases e. g. in case of storage of final products in transloading plants. Therefore the various attempts to develop a formula for calculation of the discharge quantity have only little importance for practical purposes. Sofar it has not yet been possible to develop a generally valid formula but together with the results of various experiments, these attempts give an idea of the complex problems of bulk material flow, as the flow rate describes the total flow process and therefore is an important factor for comparison. Considering experiments and their boundary conditions, it could perhaps be possible to find a general formula or at least define exactly the validity range of any such formulas.

Among several other authors, Schwedes (267) has tried to investigate the range of applicability of all existing formulas for the determination of flow rates, to transform them to a common system of measuring, and to show their different results by calculating examples. Without mentioning individual equations, his results will be discussed briefly. He differentiates between four groups of formulas: empirical, dimension analytical, energetic, and analogous to fluid flow. The various relationships appearing in these formulas show how different the boundary conditions were chosen. A primary goal was not to establish exact equations but principal relationships. Therefore important factors like bottom slope, cohesion, particle shape etc. influencing the flow rate were omitted in several instances. Schwedes tried to establish a comparison by using a general dimension-analytical equation:

$$Fr^{0.5} = \frac{\dot{M}}{g^{0.5} \cdot \varrho \cdot (d - k \cdot d_*)^{2.5}} = f\left(\varepsilon, \frac{d_*}{d}, \frac{d}{D}, \frac{D}{H}, \vartheta, \mu, \mu_*, Z\right)$$

Fr	= Froude number	ε	= porosity
$\dot{M}$	= flow rate	D	= bin diameter
g	= gravity acceleration	H	= bin height
d	= outlet diameter	ϑ	= hopper inclination

k = reducing factor
d_* = particle diameter
ϱ = specific density of material
μ = angle of internal friction
μ_* = angle of wall friction
Z = factor for particle shape

Since inertia and gravity forces are present he operates with the Froude number as the characteristic factor. The evaluation of different members can vary appreciably. Equally large are the deviations in flow rate determined by calculating the formulas with boundary conditions set by him. No uniform tendency to greater or smaller flow rates could be established, when for example the hopper slope was changed from 0° to 60°.

These equations, therefore, are valid only for the given boundary conditions and ways to transfer results to large scale, plant size units have not been established or examined. Schwedes (267) mentions the following generally valid results:

a) The proportionality between flow rate and outlet diameter can be described approximately by $M \approx d^{2,5}$, if a reduced diameter $d' = (d - k \cdot d_*)$ is introduced. The reducing factor k results from the fact that no particles flow near the outlet periphery (see also Chapter 2.2.2.).
b) Material discharge is independent of filling height H and bunker diameter D if certain values of H/D and D/d are exceeded.

In order to be able to examine the accuracy of a formula for given conditions it is reasonable to estimate the specific dependency of individual parameters with the help of model experiments. For example Matthée (194) has compared the experimental results of Reisner (241) with theoretical values derived from equations of Keneman (149), (150), and Newton, Dunham, Simpson (211). While glass pellets showed quite good confirmity with theoretical results (Fig. 7 a), gritstone samples

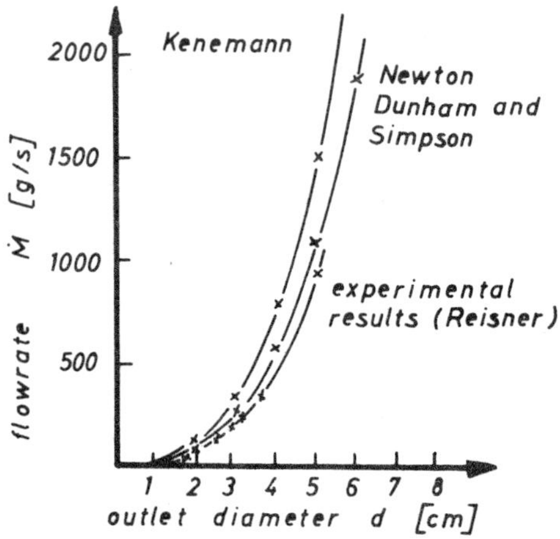

Fig. 7 a

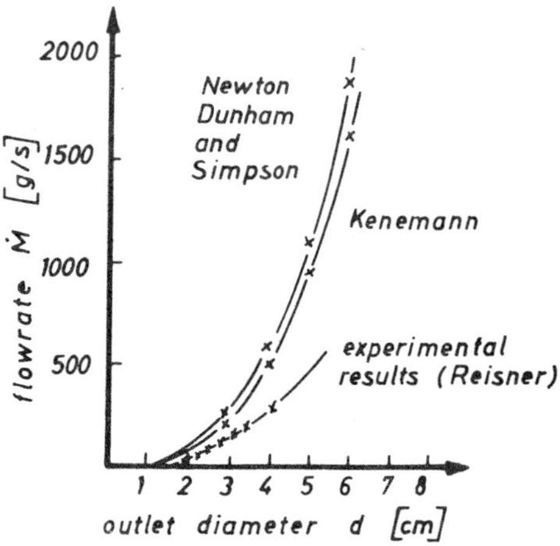

Fig. 7 b

deviated strongly (Fig. 7 b). This can also be explained by the fact that particle shape was not considered in the equations. In fundamental investigations, without considering the theoretical formulas, Reisner (241) could prove an extremely marked influence of particle shape (see Fig. 23), especially for materials having practically identical specific gravity.

1.3. Classification

In the preceeding chapters it was shown that the gravity flow of bulk materials cannot be described or calculated uniformly. On the other hand, in order to solve special engineering problems, it is necessary to estimate the validity of different results presented in the literature. Therefore means of comparison must be found by defining generally valid criteria characterizing material properties.

A rough classification should start with the terms "coarse" and "fine" or "powdered" materials, because there are basic differences (305). Only fine materials alone or in a mixture with coarser ones show cohesive properties and tend to cohesive arching. Only coarse materials can cause a mechanical interlocking. As thouroughly proved design methods exist for powders (Chapter 2.1.), the separating line, for all purposes, should be drawn at that grain size, which still can be tested in suitable shear cells. Of course this does not mean that the border line between cohesive and non-cohesive materials is defined by this value. Valentin (305) proposes that the dividing line lies somewhere between 10 and 100 mm. For the normally used test apparatus these values seem to be too high.

Kvapil (166) classifies coarse bulk materials according to their mobilities i. e. their abilities to carry out a secondary movement (Chapter 1.1. and 4.). Here the particle shape is the classifying parameter. In this connection it is also referred to the relationship between internal friction and particle shape, so that it is also possible to classify according to friction values. Examples are given for individual groups only, but not a standard method to classify these groups. As the flow of powders is determined by their cohesive properties, Jenike (133) classifies these materials according to the limiting value of their flow function FF defined by him (see Chapter 2.1.2.). These values can be estimated by standard tests.

Table 1

$FF < 2$	very cohesive and non-flowing
$2 < FF < 4$	cohesive
$4 < FF < 10$	easy flowing
$10 < FF$	free flowing

Among others these criteria underlie the classification system for bulk materials by Carr (64), (65), (66). In the classification presented in

Table 2: Classification system for masses of particles, based on average relative flow properties

description of class	unif. coef.	powder fract., %	flow-ability score	flood-ability score
I. corpuscular				
A. granules 1. uniform	1 - 10	< 5	70 - 100	0
2. nonuniform	15 - 30	< 5	60 - 75	0
3. very nonuniform	30 +	< 5	50 - 70	0
4. damp	30 +	< 5	30 - 65	0
5. soft (or sticky)	-	< 5	40 - 75	0
B. powdered granules 1. uniform	1 - 12	< 30	70 - 80	0
2. less uniform	8 - 18	< 30	55 - 70	0
3. nonuniform	15 - 30	< 30	50 - 65	0
4. very nonuniform	30 +	< 30	30 - 60	0
5. damp	30 +	< 30	40 - 65	0
6. soft (or sticky)	-	< 30	50 - 70	0
C. fluid powders 1. very a) granular (some)	-	60 - 90	55 - 70	70 - 90
b) gran. & powdr.	-	30 - 60	50 - 65	30 - 70
c) powder	-	95 - 100	45 - 65	80 - 95
2. fluid a) granular (some)	-	60 - 90	45 - 65	50 - 75
b) gran. & powdr.	-	30 - 60	40 - 60	20 - 60
c) powder	-	95 - 100	35 - 50	45 - 75
3. fluid cohesive a) granular (some)	-	60 - 90	20 - 45	25 - 60
b) gran. & powdr.	-	29 - 60	20 - 40	10 - 35
c) powder	-	95 - 100	10 - 40	15 - 45
D. cohesive powders 1. powder	-	95 - 100	5 - 25	0 - 20
2. gran. & powdr.	-	30 - 60	5 - 40	0 - 20

			avg. size range		
II. laminar					
A. micaceous	1.	thin	1/2 to 3 in.	1 - 15	0
	2.	thick	1/4 to 1 in.	10 - 20	0
	3.	powdered	-200 mesh	30 - 50	35 - 60
B. film	1.	very thin	1/16 to 1/2 in.	1 - 10	0
C. chips	1.	fine, uniform	+ 100 to + 10 mesh	70 - 80	0
	2.	fine, nonuniform	-60 to + 10 mesh	60 - 70	0
	3.	large, nonuniform	-10 to 3 in.	30 - 40	0
D. flakes	1.	thin	- 100 to 1 in.	1 - 20	0
	2.	fine, uniform	-20 to 3/8 in.	50 - 80	0
	3.	fine, nonuniform	-40 to 3/4 in.	45 - 50	0
	4.	powdered	- 200 to 1/4 in.	50 - 70	0
III. fibrillar					
A. stems	1.	very short	1/4 to 3/8 in.	50 - 60	0
	2.	short	1/4 to 1 in.	10 - 20	0
	3.	long	2 to 3 in.	1 - 10	0
B. fibrous	1.	bunches	1/2 to 2 in.	5 - 20	0
	2.	fine bunches	- 100 mesh	20 - 30	0
	3.	powder, coarse	- 200 to + 10 mesh	40 - 55	0
C. acicular	1.	fine	+ 100 to + 40 mesh	65 - 75	0
	2.	fine, nonuniform	- 100 to + 10 mesh	55 - 70	0
	3.	medium	+ 20 to 7/8 in.	30 - 40	0
	4.	powder, fluid	- 200 mesh	35 - 50	40 - 60

Table 2 the particle shape is to be regarded as a parameter of first degree. Criteria for the sub-classes and further subdivisions are size, available surface area and the effective surface forces on the surface area. The strict classification between coarse and fine material which quite often is not possible, can be modified by considering the uniformity coefficient, the part of fine material, and the average size range of the bulk material in special columns. The different classes are characterized by numerical values of flowability and floodability of bulk materials. These values result from four standard tests, in which each test is evaluated with a maximum of 25 points. In Table 3 the flowability values of a standard free flowing material (sand) are compared with a very poor flowing specimen (calcium carbonate).

Table 3

flow property	sand		calcium carbonate	
	value	points	value	points
angle of repose	25°	25	50°	12
compressibility	8 %	23	32 %	9,5
angle of spatula	25°	25	93°	2
alternates:				
cohesion or	0 %	25	60 %	2
uniformity coefficient	2	23		
		98 or 96		25,5

angle of spatula: a spatula is lifted straight up out of the material,

cohesion: percentage of a sample - 200 mesh remaining after a certain time of vibrating a nest of screens,

uniformity coefficient: ratio of the width of sieve opening that will pass 60 % and the sieve opening that will just pass 10 % of the sample.

Similar standard tests determine the floodability:

a) flowability: as evaluated by the tests above,

b) angle of fall: a small weight is dropped several times on a plate on which a heap of material rests under its natural angle of repose,

c) angle of difference: difference between angle of repose and angle of fall,

d) dispersibility: loss of weight of a sample of bulk material during free fall.

Furthermore, apart from these general classifications, a properties code is established, in which all possible properties to be considered during handling of certain materials are included e. g. dustiness, corrosiveness, abrasiveness, hygroscopic nature etc. These materials are listed as follows in Table 4:

Table 4

material	bulk density (lb/cu. ft.)			class	properties code	remark
	loose	pack	work			
for example:						
potassium chloride	67	85	71	1C2b	CJPtwy	KCl impure

The properties code in this case has the following legend:
C = dusty, J = forms or can have lumps, P = highly corrosive if wet or dissolved, t = hygroscopicity makes material less flowable and floodable, w = tendency to mass, y = flow becomes poorer if packed excessively.

Such a classification has only little value as an exact scientific description, since the same evaluation of the four tests by the above mentioned point system seems to be arbitrary as long as it is not well founded. But for the plant engineer this classification certainly has its merits, as the required tests are simple and flow properties of specific materials can easily be estimated and classified. In this way it becomes possible to forecast, wether the material flows freely, tends to arch or tends to be fluidized, wether accessories such as rotors, vibrators or others are necessary, or if it needs special engineering and attention. Carr has already tested 3,500 dry materials in the above mentioned way, so that in most cases it is sufficient to consult his tables to know in advance which difficulties might have to be expected.

2. Design Based on Flow Properties of Bulk Materials

Basically there are two technically satisfactory ways of designing bunkers correctly:

a) the conventional bin-hopper system and
b) special bunker constructions for the different types of discharge equipment.

The choice between these two possibilities can be only made by comparing the costs of installation as well as that of operation. In the second case, the form of the bunker is solely determined by the corresponding method of discharge. For the first case, to be discussed in this chapter, the bunker has to be carefully constructed to suit the flow properties of the materials to be stored. The following criteria indicate when a core flow or a mass flow is required.

	mass flow	core flow
very uniform flow	required	not required
first-in/first-out	required	not required

A completely uniform flow, for example, is required for bulk materials that tend to fluidize (136), or where a first-in/first-out sequence is desired as in case of chemical reactions during storage or when segregation and degradation are to be taken into account (134). This procedure may vary for every process and must be examined very carefully, comparing the technical advantages of a mass flow hopper with the disadvantages of construction costs.

2.1. Design of Mass Flow Hoppers

In recent years the method of Jenike for design of mass flow hoppers has become standard in the mechanical processing industries. Numerous applications have confirmed its applicability. In this connection the works of Jenike and his coworkers are to be specially recommended. They contain detailed descriptions of test procedure, measurements of material properties (129), (132), an analysis of stress conditions in hoppers (132), (133), and the formulas as well as graphical results derived from them (133). A similar method has been developed by Walker (308), (309) and Shinohara (273). The calculations of stress conditions are presented by a number of authors (100), (101), (102), (135), (142), (180), (251). Detailed abstracts about this are to be found in Richards (249) and Schwedes (267). The theory of Jenike is specially discussed by Herne (118), and in a dissertation to be completed in the near future at the Technical University, Braunschweig, Germany.

A discussion of all the works mentioned above can hardly be made in the limited scope of this chapter and would certainly not lead to a better understanding. Therefore, the representation of the theoretical calculation is purposely avoided. In order to justify the importance of such methods for practical bunker design, the fundamental steps to derive material properties and to determine the critical hopper size are outlined based on the method of Jenike.

2.1.1. The Shear Cell Test

In case of mass flow the whole bulk volume is in motion after the flow has set in. To insure continued flow it is necessary that the strength of the bulk solid does not exceed the stresses in the material at any point in a would-be obstruction. To measure the strength of a bulk solid when consolidated under the same conditions of pressure, temperature and moisture content as exist in the hopper Jenike has constructed the Flow Factor Tester shown schematically in Fig. 8. This tester is essentially a further development of the shear apparatus used for measuring material strength in soil mechanics.

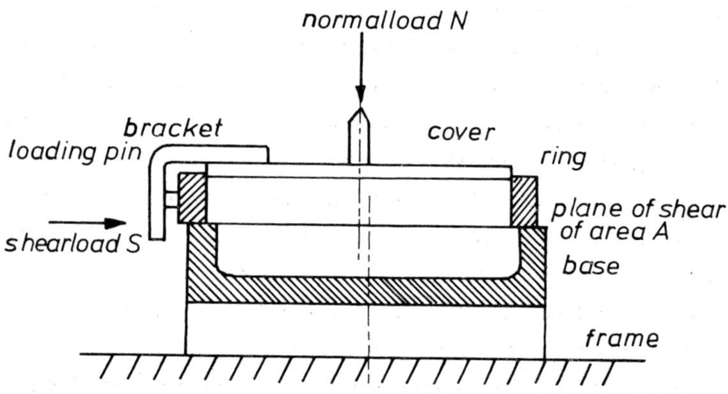

Fig. 8

The test procedure is as follows. To reproduce the consolidation inside the hopper the samples of a particular series of experiments are consolidated under shear using a constant normal load. After removing the consolidation load one sample is sheared off under a reduced normal load N and the required shearing force S is measured. This process is repeated for the other samples using different loads N each less than the consolidating load.

The forces N and S can be converted to normal stress σ_N and shearing stress τ by dividing by the area A of the shear cell. By plotting the corresponding pair of values in a σ_N, τ-diagram one gets the yield locus (Fig. 9 a). In general the yield locus is a curve, but is very often approximated by a straight line characterized by its intercept C on the τ-axis called the cohesion of the material, and the slope angle ϕ of the line. The tensile portion of the yield locus represented by the dashed line and characterized by the value of T is difficult to measure with a shear cell. Fortunately this portion of the yield locus has little practical significance designing bins and hoppers. If the consolidating load is changed a different yield locus is obtained.

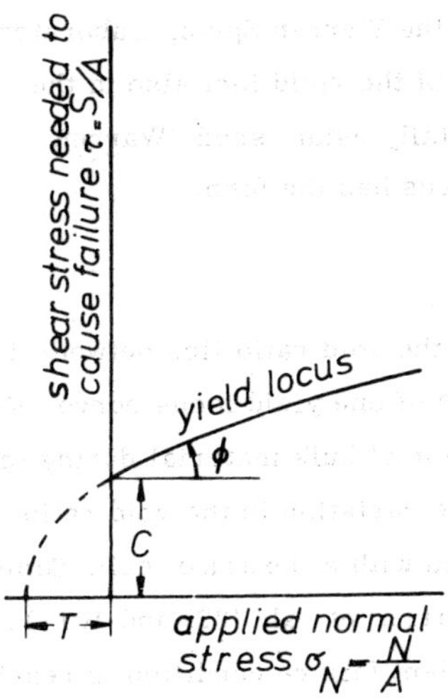

Fig. 9 a

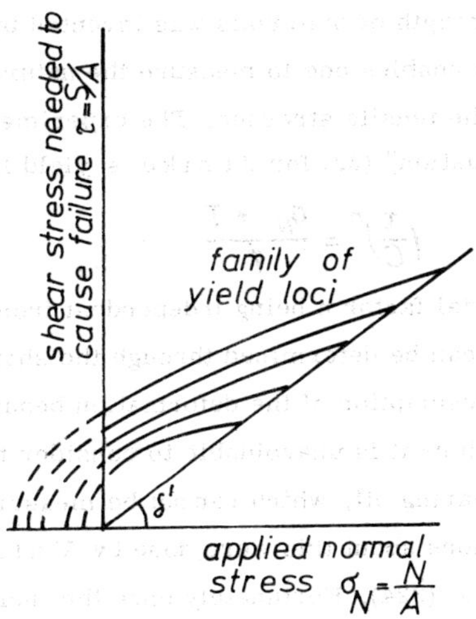

Fig. 9 b

The family of yield loci presented in Fig. 9 b is obtained when a number of tests are performed with different consolidation loads. For many bulk materials the flowability varies due to external influences such as temperature variations, change in percentage of water content or through consolidation during storage at rest. For these cases a series of analogous tests can be performed in which the samples are subjected to such influences. The line connecting the end points of the yield loci is practically a straight line. The angle δ' characterizes the failure conditions for various states of preconsolidation.

The proposals of Jenike to measure the flow properties of powders have been specially taken up and developed further in England. Williams and Birks (322) have investigated as to how far the preparation of the powder specimens influences the result of the shear cell test and have tried to simplify and standardize the test procedure. An apparatus to measure the tensile strength of materials was invented by the Warren Spring Laboratory (21) which enables one to measure the nature of the yield loci also in the range of the tensile stresses. The experimentally established "Warren Spring Equation" (22) for Jenike's yield locus has the form

$$\left(\frac{\tau}{C}\right)^n = \frac{\sigma_N + T}{T}$$

The material factor n being independent from the void ratio lies between 1 and 2 and can be determined through the shape of one yield locus curve. For an exact description of the deformation behavior of bulk material during incipient failure it is unavoidable to consider the variation in the void ratio during shearing off, which can not be measured with a Jenike cell. Some investigations about this were done by Valentin et al. (22) and Williams and Birks (324). Fortunately once the incipient failure condition is reached the state of stress rapidly approaches a steady flow condition described by Jenike's Effective Yield Locus.

2.1.2. Determination of Necessary Data of Bulk Materials

For design of mass flow hoppers according to Jenike's method a number of characteristics of the bulk materials are required, which can be derived from the shear tests.

a) Unconfined yield strength f_c and flow function FF

As shown in Fig. 9 b the strength of bulk materials are very much dependent on the preconsolidation. In order to find out the conditions under which a stable arch can build at the outlet, the strength of the arch i. e. strength of the material at a free surface must be determined. Jenike (134) has illustrated this fact with the following thoughts.

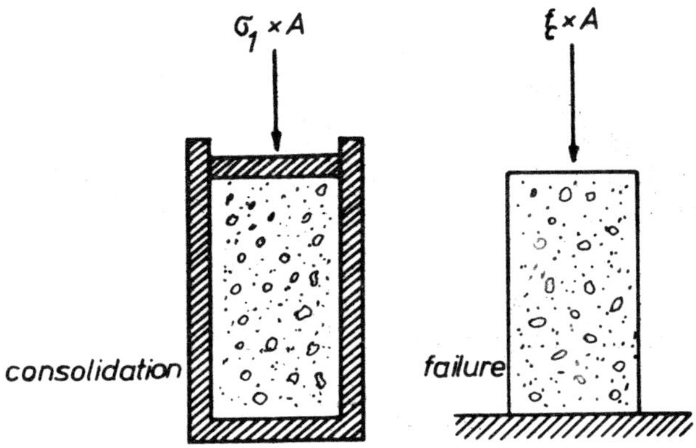

Fig. 10

A powder is consolidated in an idealized cylinder with frictionless walls under a definite pressure σ_1; the cylinder is then removed and the cylindrical sample is loaded without any lateral support. The pressure f_c necessary to destroy the sample is equal to the surface strength of the sample for a given state of preconsolidation. The values of f_c and corresponding consolidation stress σ_1 can be determined from the yield locus diagram as shown in Fig. 11. The stress conditions in the bulk material during shearing failure are represented by Mohr circles tangent to the yield loci. Because at a free surface the stress normal to the surface as well as the shear stress in the surface are zero, f_c is given by the M o h r circle which is tangent to the yield locus and passes through the origin. The corresponding value σ_1 for the consolidation stress is given by the M o h r circle through the end point of the yield locus.

The flow function is defined by the relationship

$$FF = \sigma_1 / f_c$$

The graphical representation of FF is shown in Figure 11.

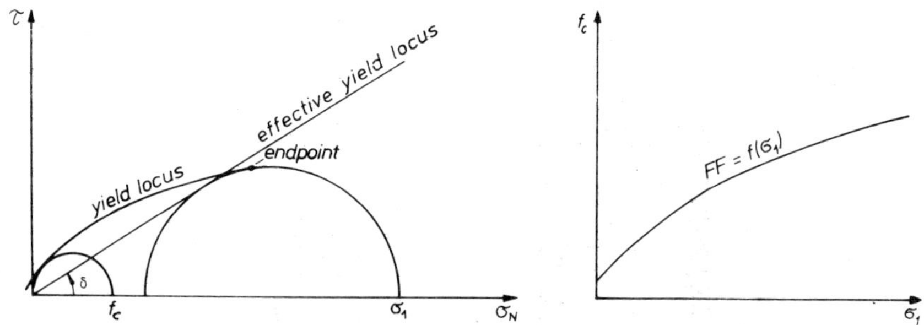

Fig. 11

At the University of Bradford the so called "confined failure apparatus" was invented which permits the direct measurement of f_c and σ_1 to avoid the inaccuracy in the graphical determination of f_c (45). Here the uniaxial compression test was performed similar to that described in Fig. 10 in which the friction between the sample and the wall during consolidation was practically nullified. Results from these tests showed good agreement with tests on the same material in the J e n i k e shear cell (12).

b) Effective yield locus EYL and effective angle of internal friction δ

In Chapter 2. 1. 1. it was mentioned that the curve through the end points of the yield loci reproduces the conditions in a flowing powder. To derive the stress conditions in hoppers Jenike has used the effective yield locus which is tangent to the largest Mohr circles i. e. tangent to the Mohr circles passing through the end point of yield loci.

The angle δ' and the effective angle of internal friction δ, which is also practically constant, vary about 5° (12). The effective yield locus is

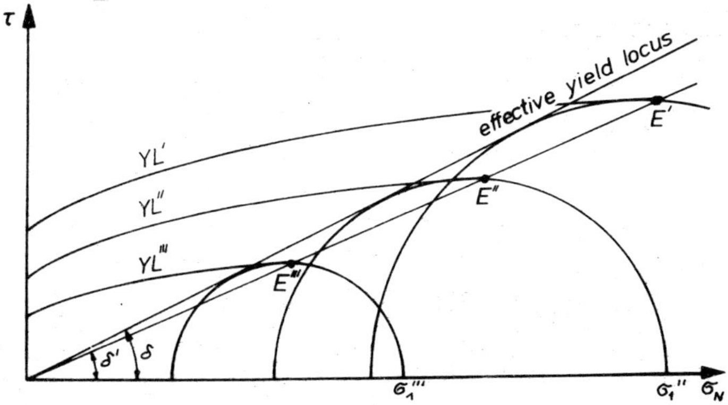

Fig. 12

advantageous in mathematical treatment of the problems, because it gives a simple relationship for the principal stresses σ_1 and σ_2 :

$$\sigma_1/\sigma_2 = \frac{1+\sin\delta}{1-\sin\delta}$$

c) Wall yiels locus WYL and angle of wall friction ϕ'

To determine the angle of friction between solid and wall the same Flow Factor Tester can be applied. The shear stresses necessary for different normal loads to move the sample of the solid over the sample of the wall material are measured. Plotting σ_N vs. τ a wall yield locus is obtained, which is usually a straight line through the origin. However, occasionally a curved wall yield locus is obtained and ϕ' is determined as shown in Fig. 13 as a function of normal pressure.

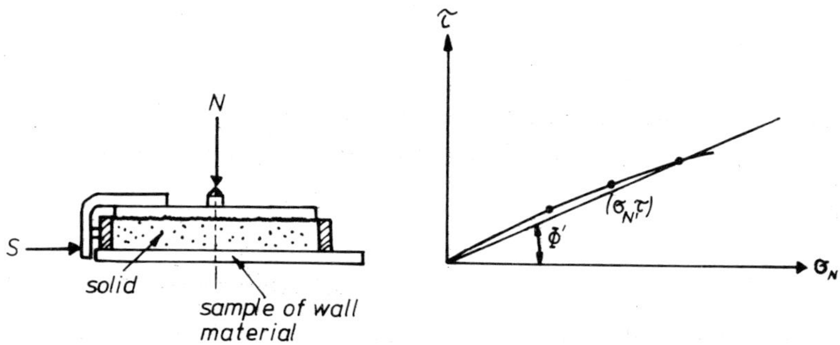

Fig. 13

2.1.3. Critical Hopper Dimensions

In order to characterize the flow properties of a hopper Jenike defines the flow factor ff = $\sigma_1 / \bar{\sigma}_2$ as the ratio between consolidation stress σ_1 and the stress $\bar{\sigma}_1$ which acts on an obstruction.

With lower values of ff the hopper flow conditions are better. For a given hopper ff is constant, because in the hopper region σ_1 and $\bar{\sigma}_1$ are linearly dependent on hopper diameter. From theoretical analysis of stress conditions the following relationship is established:

$$ff = (1+m)\,\frac{S(\Theta')(1+\sin\delta)}{2\sin\Theta'}$$

m = 1 for axial symmetry (conical outlet)

m = 0 for plane symmetry (slot outlet)

r, Θ = polar coordinates

S = function of Θ'

Θ' = hopper slope measured from the vertical

δ = effective angle of friction

Values of ff have been calculated - for various boundary conditions, effective angles of internal friction δ, angles of wall friction ϕ' and angles of hopper inclination Θ'. Typical results for δ = 40° are shown in Fig. 14.

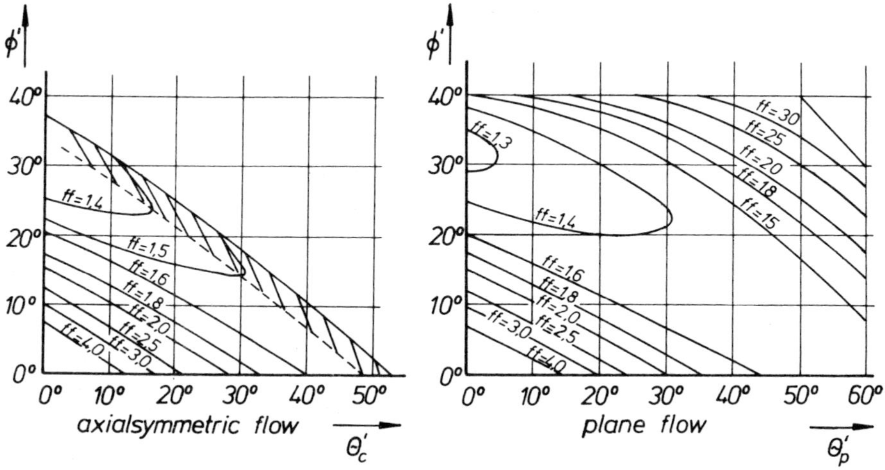

Fig. 14

Beyond the zone in which the ff-curves are plotted no solutions of the equations exist and thus no mass flow can take place. From the equation for ff and from the charts it is evident that in case of plane symmetry the flow conditions of a hopper are more favorable than in case of axial symmetry.

From these charts the hopper slope is first determined. For given δ and ϕ' the value of hopper inclination can be selected arbitrarily within the mass flow region. For practical reasons it is chosen as large as possible in order to keep the construction expenditures low. However, it is necessary to stay within the boundary line by some degree to compensate for the errors arising eventually during performance and graphical evaluation of the shear tests. With selection of hopper slope the flow factor for this system is fixed. The formula for the minimum outlet diameter d is

$$d = \frac{H(\Theta') \cdot f_{c\,crit}}{\gamma}$$

The relation for H (Θ') is shown in Fig. 15. Here also it is evident that for plane flow considerably smaller outlet diameters can be chosen.

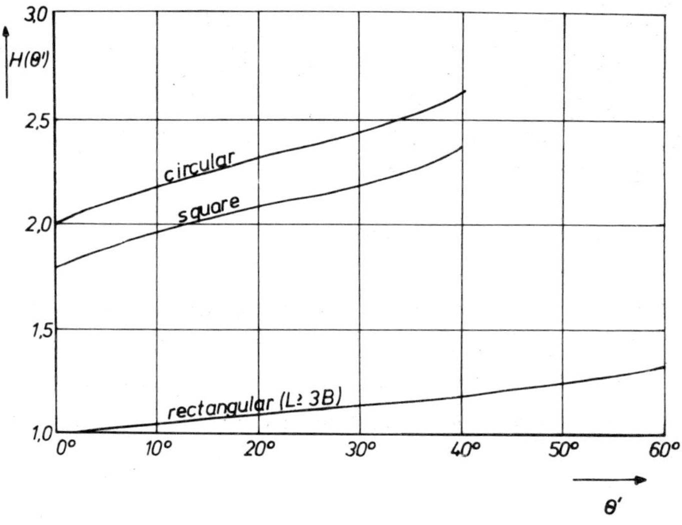

Fig. 15

f_c is not constant but dependent on the consolidation stress σ_1 (see Fig. 11 and 16). Arching is only possible when the strength of material f_c is greater than or equal to the stress $\bar{\sigma}_1$ acting in the arch. The critical value for f_c is therefore given by the point of intersection of the flow function and the flow factor. Above this point the conditions for mass flow are fulfilled.

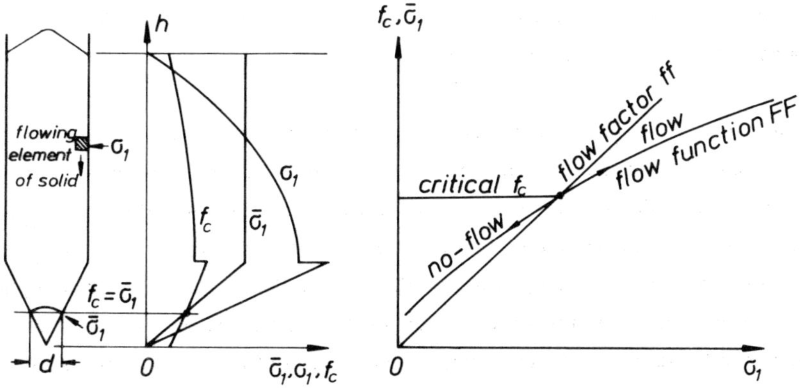

Fig. 16

Concludingly the following steps are necessary to determine the critical hopper dimensions:

a) performance of the shear test, plotting of the yield loci in a σ_N - τ - diagram, determination of the effective angle of internal friction δ and the angle of wall friction ϕ',

b) for the values of δ and ϕ' the angle of hopper inclination Θ' is selected within the mass flow region from the charts for plane or axial symmetry; with Θ' the flow factor ff for this hopper is fixed,

c) determination of the values of f_c and σ_1 from the corresponding Mohr circles and plotting the flow function FF = σ_1 / f_c over the flow factor ff.

d) calculation of the minimum outlet diameter with f_c crit. and H (Θ').

2.1.4. Range of Application

The method of Jenike is principially applicable to fine grained materials. During the past five years, a number of publications reviewed the performance of material tests and the design of hoppers based on such tests, for example for bituminous coal (72), iron ore concentrate (134), (286), ground phosphate (270), cement (265), etc.

The range of grain sizes that can be tested is limited by the dimensions of the shear cell, about 95.3 mm inside diameter. As a general rule, Jenike assumes that the solids which do not contain particles smaller than 0.25 mm are free-flowing. This assumption is not everywhere supported. Valentin, for example, has set the grain size limit more than a decimal power higher to 10 mm (305). For the Flow Factor Tester, the max. particle size is arbitrarily fixed at 0.84 mm, being about 1/100 of the shear cell diameter.

In many instances materials to be handled have a particle size exceeding the limiting value and thus cannot be tested in a shear cell. According to Jenike (132), (133), the properties of fine particles alone are decisive for the flow properties. "The coarse ones are a passive agent and do not develop yield strength without fines to bind them." It suffices to take the shear characteristics of fine materials under 0.84 mm as the basis for the hopper designs. Sofar it has not been investigated if the shear strength of a material is lowered by coarse particles, which would lead to a more favourable slope of the hopper. It is reasonable to assume that a larger particle would create a weak point in a uniform layer of fine material. Furthermore the minimum percentage of fine particles should be properly defined in order to make Jenike's statement valid. Further investigations are required to establish if the percentage of fine material is high enough to define a bulk solid as non free-flowing or not, and whether it is justified to select a more expensive method of mass-flow design. Some suggestions for classification are to be found in the system proposed by Carr (64), (65), (66), (see Chapter 1.3.).

Jenike (133) classifies powdered materials according to the shape of their flow function (Chapter 1. 3.). The parameter for classification is the limiting FF-value. Bulk materials with values of FF < 2 are described as very cohesive and non-flowing. For these materials a free discharge under the influence of gravity only is not any longer possible i. e. the outlet dimensions are not recommended for economical construction (268). In such cases, other methods of discharge should to be considered.

When first-in/first-out or a very uniform flow is required, the bunkers for coarse materials are also to be designed according to the mass-flow criterion. In case very free-flowing materials are concerned, a minimum of difference between the particle velocity in center and at the wall is desired. Suitable design methods for these cases are not available using theoretical techniques so that experimental methods are required. Some information about this is presented by Kvapil et al. (Chapter 1. 1.). The determination of the minimum outlet diameter necessary to avoid an arching is discussed in chapter 2. 2. 2.

2. 2. Design of Core Flow Bins

A core flow pattern is characterized by the formation of zones of different mobility in the bunker (Chapter 1. 1.). In order to maintain a continuous flow and to utilize the whole bunker volume, i. e. to discharge also the passive zones which are outside the flow channel, the outlet diameter as well as the hopper inclination must be determined in accordance with the bulk material properties.

2. 2. 1. Powdered Materials

In the case of fine grained cohesive material it is possible that only the zone above the outlet is discharged and subsequently a pipe is formed, leading to a total flow stoppage. Similar to the mass-flow criteria, Johanson has calculated the boundary conditions at which stable pipes cannot occur (142). For this calculation, the fundamental material tests described in Chapter 2.1.1. are required. The need of a time-consolidation test must be carefully examined. When the bunker is filled and discharged simultaneously, the external zones may possibly remain at rest for a long time. Then, the time yield locus and time flow function are to be determined. Johanson expressly refers to the fact, that the method of determining the flow factor for piping as described by Jenike (133) is valid for specific flow conditions only. Its general application may lead to wrong results. Latest investigations have shown, that during bunker charging remarkably larger vertical consolidation stresses occur than during steady flow of the material (Chapter 3). In designing a bunker, the most unfavourable case must be taken into consideration and, therefore, the critical flow factor must be determined by the initial pressures during filling. Johanson has established formulas for the calculation of flow factors for cylindrical as well as conical bins. Taking the conical hopper as an example, the method to determine the critical outlet diameter, which avoids piping, should be briefly noted here. The flow factor is defined in Chapter 2.1.3. as

$$ff = \sigma_1 / \bar{\sigma}_1$$

σ_1 can be calculated from

$$\sigma_1 = \gamma \cdot B \cdot F$$

where B is the diameter (or width) of the hopper at any level and F is a function of bin geometry and that of k, which is defined in Fig. 17. A typical plot for the consolidating pressure factor F (for k = 0,8 and m = 1) is shown in Fig. 17.

$\bar{\sigma}_1$ can be calculated from

$$\bar{\sigma}_1 = \frac{\gamma \cdot d}{G(\phi)}$$

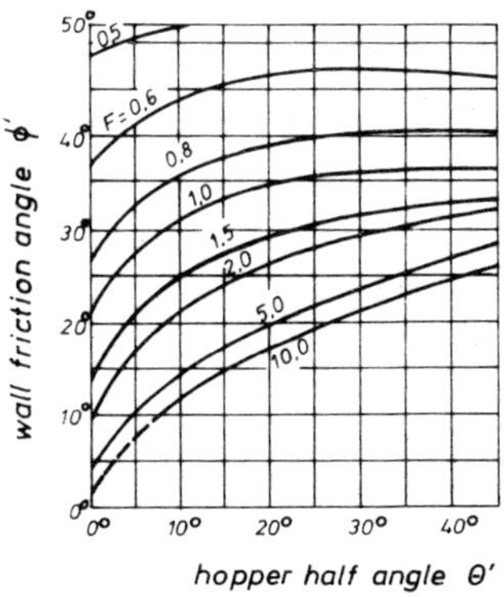

$$k = \frac{1 - \nu(\alpha + m)}{1 + \alpha - \nu(\alpha - m)}$$

ν = Poisson's ratio

α = coefficient of compressibility of a solid

m = coefficient:
$m=0$ for long rectangular bins
$m=1$ for circular bins

Fig. 17

where d is the width of the flow channel and G (ϕ) the ratholing factor which can be deduced from Fig. 18. A new factor, the static angle of internal friction ϕ (respectively ϕ_t for time consolidation) developed within the solid

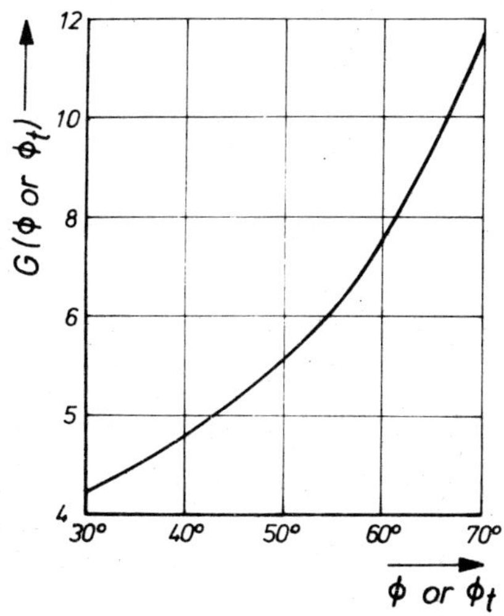

Fig. 18

at an exposed surface of a pipe has to be introduced. It can be determined from a σ_N - τ - diagram as the inclination of the yield locus at a point where the Mohr circle through the origin touches the yield locus or time yield locus (Fig. 19).

The unconfined yield strength f_c or time yield strength f_{ct} as well as the flow function FF of the materials were already determined by these Mohr circles (Chapter 2.1.2.).

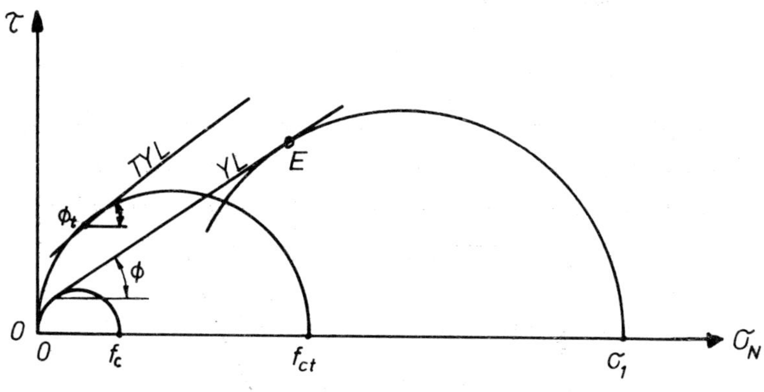

Fig. 19

The next step to calculate the minimum outlet width for avoiding piping is to draw the flow factor over the flow function. The point of intersection gives the critical value for f_c, needed in the formula

$$d = \frac{f_{c\,crit}}{\gamma} \cdot G(\Phi)$$

At this stage it must be noted, that the minimum hopper slope angle with the horizontal (90° - Θ') has to be selected greater than the angle of wall friction of the material, so that the bulk material can be discharged completely out of the bin.

The steps to calculate the flow factors in the case of flat bottom silos are similar. For σ_1 Johanson has only substituted the average filling pressure calculated from the Janssen formula according to the following equation (see Chapter 3.)

$$\sigma_1 = \frac{\gamma \cdot R}{K \cdot \tan \phi'} \left[1 - e^{-4K \cdot \tan \phi' \cdot h/D} \right] \quad ^{x)}$$

where

R = hydraulic radius F/U = D/4 (for a round bunker),

D = diameter of the bunker,

h = height of the bunker, measured from the surface of the bulk material.

In contradiction to Janssen/Koenen (Chapter 3.) K is calculated with the help of effective angle of internal friction δ

$$K = \frac{1 - \sin\delta}{1 + \sin\delta}$$

For very shallow silos with a ratio $0 < h/D < 1$ the initial consolidation pressure can be assumed simply as

$$\sigma_1 = \gamma \cdot h$$

By the combination of σ_1 and $\bar{\sigma}_1$ the flow factor ff for piping and thus the critical diameter can be calculated.

2. 2. 2. Coarse Materials

Coarse materials are basically free flowing i. e. under consolidating pressure they do not develop much of an unconfined yield-strength. The yield locus calculated under the conditions specified in Chapter 2. 1. 1. would begin in origin, resulting in $f_c = 0$ (Fig. 20). The arching and piping due to

----

x) The expression in the brackets in the above equation is by mistake taken to be $[1 - e^{4K \cdot \tan \phi' \cdot h/D}]$ by Johanson (142). This would lead to negative values of σ_1.

material strength are thus not possible.

The critical diameter, at which no flow of material occurs, does not have to be hundred or thousand times the max. particle size, but it lies within

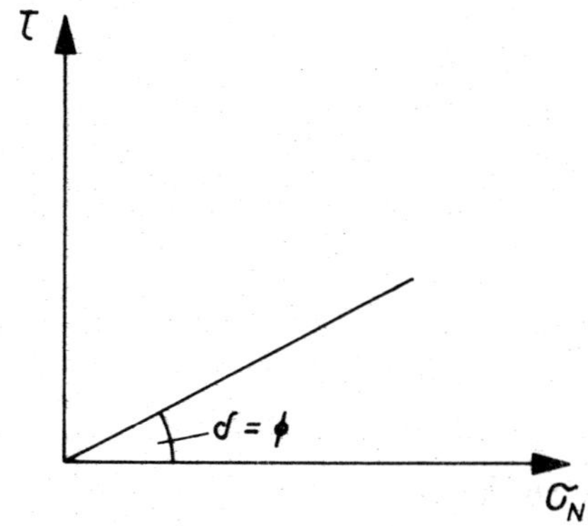

Fig. 20

the range of a few times the max. particle size. To determine the equilibrium conditions of such a particle arrangement, referred to as microstructure by Peschl (225), the exterior forces and the properties of individual particles must be taken into account. The consideration of average tension in the bulk material as well as the bulk behavior of particles, as it is practiced in the field of soil mechanics, is not permissible. As an example Fig. 21 (166) shows the force transmission in a photoelastic model of a mechanical arch.

Peschl (224),(225) has occupied himself very intensively with the theory of arching in coarse materials. His investigations have shown that a theoretical analysis of stresses alone does not lead to the determination of

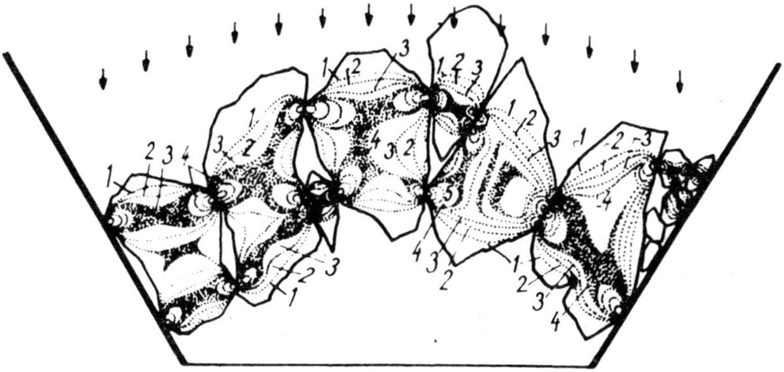

Fig. 21

boundary values for d/d_* = outlet diameter/maximum particle diameter. The statistic character of arching is for example illustrated in Fig. 22 (225), in which the different forms of possible arches and their asymmetry

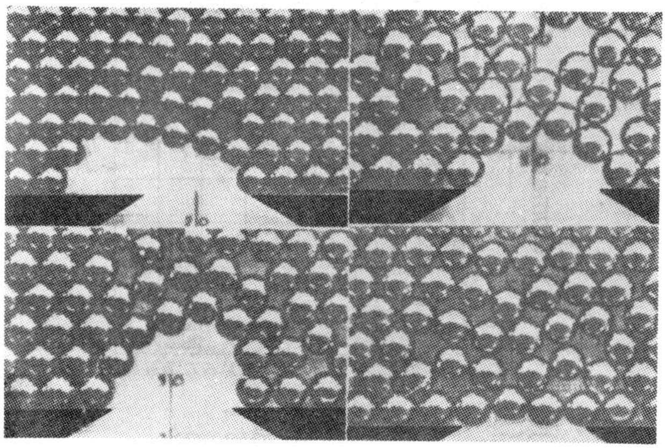

Fig. 22

is presented. For the unfavorable case the following relationship is established

$$(d/d_*)_{max} = 2 + \frac{\alpha + \varrho_{wm}}{\varrho_{mm} \cdot C(\varrho_{mm})}$$

α = hopper slope

ϱ_{wm} = angle of wall friction

ϱ_{mm} = angle of internal friction

For small values of ϱ_{mm}: $\varrho_{mm} \cdot c(\varrho_{mm}) / \varrho_{wm} = 1/2$

For large values of ϱ_{mm}: $\varrho_{mm} \cdot c(\varrho_{mm}) / \varrho_{wm} = 1$

Because the hopper slope is included in the equation, an optimum hopper form can be derived from the above equation. From Reisner's

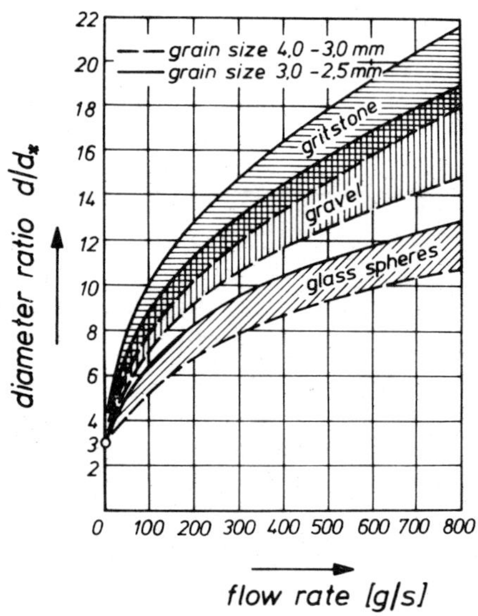

Fig. 23

diagram (241) (Fig. 23) showing the relationship between flow rate $\dot{M}$ and the ratio d/d_* , the theoretical limit reaches a value of 3, which is also confirmed by the results of Takahashi (287).

Brown and Richards (58) explain the mechanism of arching through a series of experimental tests. They have found that there is an empty space adjacent to the edge of the aperture within which practically no flow can take place (Fig. 24).

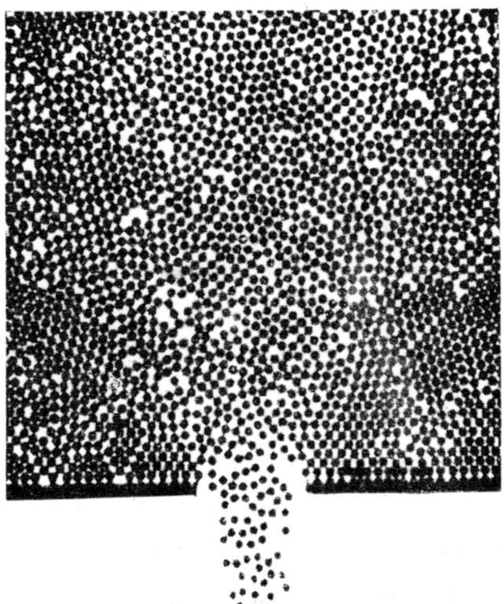

Fig. 24

Its width of k/2 is a function of material properties alone. From these experiments it was obvious that the critical diameter for a slot discharge corresponds to a value of k, whereas in the case of circular discharge it amounts to 2 k. Analogous to the theory of Jenike, it also shows in case of coarse materials that a remarkably better flow condition exists for a plane flow.

These theoretical investigations can be applied to practical cases only in a limited way, because the reference to real materials is missing i. e. the value of k in its dependence upon material properties as grain size or particle shape is not defined specifically by Brown and Richards (58). Because of this, two empirical formulas are presented below, which take the grain characteristics into account. Kvapil (167), (169) starts from a minimum value of 5 for the ratio of d/d_m (d_m = average grain size). For square or circular cross-section of the outlet it results in

$$F_s = (5 d_m)^2 \cdot k \qquad \text{or} \qquad b_s / d_m = 5\sqrt{k}$$

$$F_c = (5 d_m)^2 \cdot k \qquad \text{or} \qquad d_c / d_m = 5\sqrt{1.08 \cdot k}$$

F = outlet area — index s = square

b = length of side — index c = circular

k is a correction factor which expresses the grain distribution as well as the percentage of adhesive constituents. The k-value can be determined approximately from Fig. 25 (167), following the direction as indicated by the arrows. Region I characterizes the percentage and the shape of coarse lumps, region II and III identify the percentage of middle and fine material and region IV the adhesive constituents.

In the case of the adhesive contents being more than 10 %, the value of k is uncertain. It must generally be investigated, if cohesive arching occurs and if accordingly larger diameters are necessary.

Langmaid and Rose (176) supply two empirical formulas for circular and slot discharge apertures and refer to the average grain diameter d_m, too. Under the presumption of a narrow grain distribution the critical ratio for slotted outlet becomes

$$b/d_m = 1.8 + 0.038\,(f/k)^{1.8}$$

and for circular outlets

$$d/d_m = 2.3 + 0.071\,(f/k)^{1.8}$$

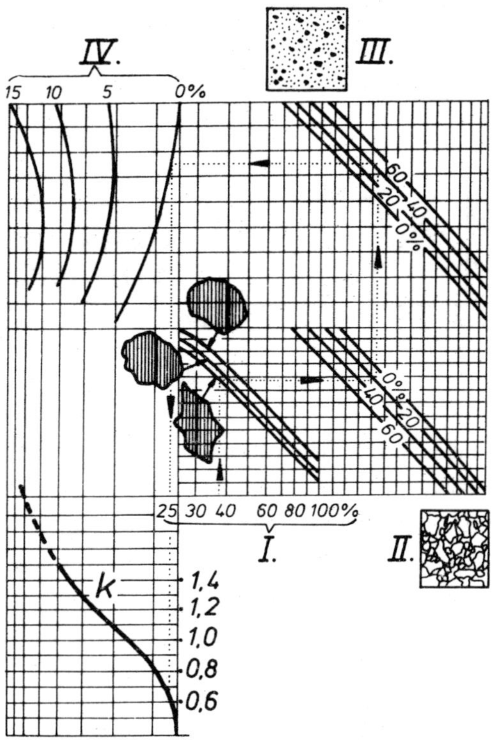

Fig. 25

The second term in the above equation includes the grain shape of the material, where, according to H e y w o o d (119), f means the surface coefficient and k the volume coefficient. For coal for example the ratio f/k lies within the magnitude of 10 (111) so that

$$b/d_m = 4.2 \quad \text{and} \quad d/d_m = 6.8$$

The investigation of the statistic character of outflow behaviour of coarse material, according to W ö h l b i e r and R e i s n e r (325), A y t a m a n (25) and P e s c h l (224), (225) lead to the fact, that, in addition to the no flow range, two more flow ranges are identified. These are the probable and the free flow range. The probable flow range can be characterized by the frequency of stable arches. In Fig. 26 (225) P e s c h l proves that the limiting

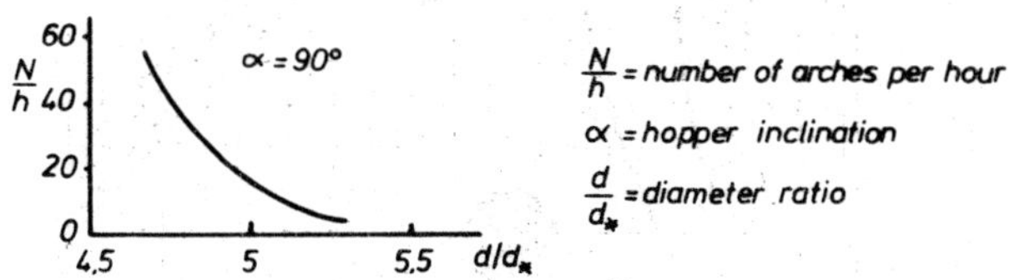

Fig. 26

value d/d_* for the probable flow range cannot be determined exactly with help of experiments. The frequency of the formation of stable arches is plotted against the diameter ratio for a flat bottomed bin ($\alpha = 90^o$). The transition to the free flow range is characterized by the formation of unstable or dynamical arches. Presented in dimensionless factors

$$\eta = \frac{v}{\sqrt{g \cdot d}} \quad \text{or} \quad \eta = \frac{v \cdot \alpha^{0.52}}{\sqrt{g \cdot d}}$$

and

$$\delta = d/d_*$$

P e s c h l calculated the velocity function of the outflowing material shown in Fig. 27 (225). In the transition zone, the discharge velocity does not increase due to the increase in the diameter ratio alone, but also because of the lower frequency and duration of the dynamic arches. Only above a certain limiting value d/d_* is the discharge velocity v solely a function of the diameter ratio.

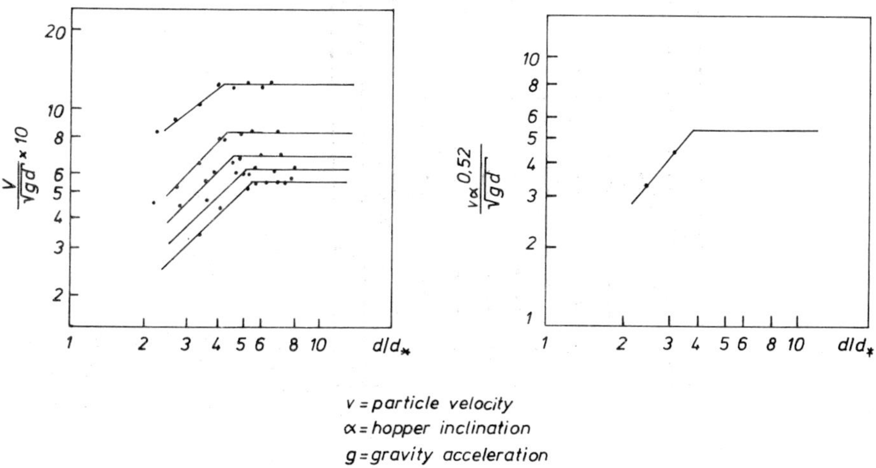

Fig. 27

Similar to the determination of the flow rate, no generally valid formula is established for the determination of the critical diameter. The influencing factors and the experimental boundary conditions are so diverse, that it is not yet possible to describe the bulk material as a whole with a characteristic factor, as for example the effective angle of internal friction defined by J e n i k e. According to S c h w e d e s (268), it can be assumed, that, in order to guarantee a continuous flow, the diameter ratio d/d_* for coarse materials must be at least 10.

3. Loads of Bins and Bunkers

The classical silo theory with the formula for the filling condition has been established by Janssen and Koenen (128), (155), (Fig. 28).

$$F \cdot p_V + \gamma \cdot F \cdot dz = F(p_V + dp_V) + p_W \cdot U \cdot dz$$

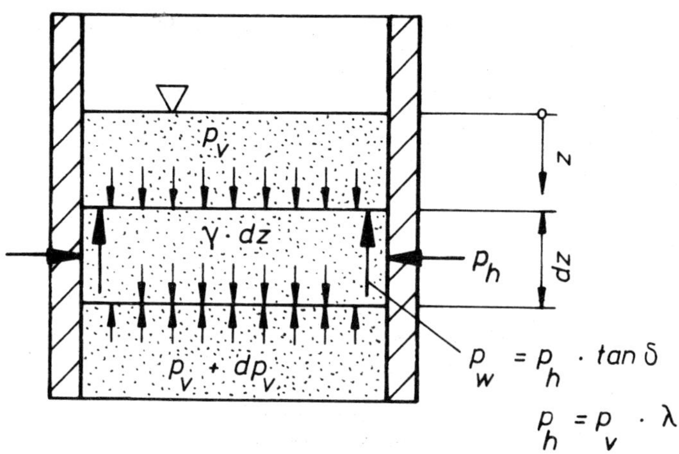

Fig. 28

The solutions for this are as follows:

$$p_v = \frac{\gamma \cdot F}{\tan\delta \cdot \lambda \cdot U} \cdot A$$

where

$$A = (1 - e^{-\tan\delta \cdot \lambda \cdot U z / F})$$

$$p_h = \frac{\gamma \cdot F}{\tan\delta \cdot U} \cdot A$$

$$p_w = \frac{\gamma \cdot F}{U} \cdot A$$

F = cross-section of the bunker

U = circumference of the bunker

γ = bulk weight of the material

$\lambda = p_h/p_v$ = const.(pressure coefficient of silo)

$\tan \delta = p_w/p_h$ = const.(angle of wall friction)

The qualitative character of the curve shows that contrary to fluids, the pressure values tend to approach a limit value. For larger z-values, A becomes 1. Recent studies (228), 240) confirm the formula of J a n s s e n and K o e n e n qualitatively. λ and $\tan \delta$ can be regarded as constant over the filling height, though their quantitative dependence from the properties of bulk materials are still uncertain.

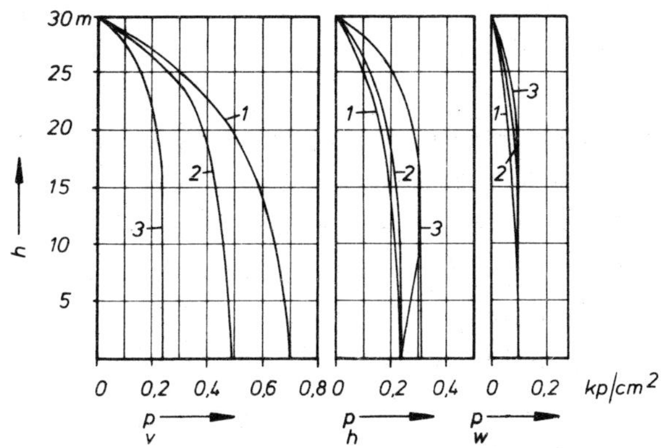

1 Janssen / Koenen

2 standard DIN 1055, page 6: filling

3 standard DIN 1055, page 6: discharge

Fig. 29

During the discharge a reorientation of the stress condition takes place. This fact became apparent when some silos broke down, which were constructed according to the above mentioned formula with a very low safety factor.

The deficiencies in the formula of Janssen and Koenen as well as discrepancies in other theoretical and experimental works are discussed in detail by Pieper and Wenzel (228).

3.1. Standards for Load Calculation

Most of the investigations about the dynamic wall pressures during the flow of bulk materials have been performed in typical core flow bins, frequently with a horizontal bottom. Jenike´s definitions of mass-flow and core-flow were not known at those times.

The fundamental experiments of Pieper and Wenzel (227), (228) have lead to new standards in Germany (8). The validity of the formula for filling pressure of Janssen and Koenen is confirmed when the silo pressure coefficients are replaced by the static pressure coefficient λ_0 commonly used in soil mechanics:

$$\lambda_0 \approx 1 - \sin\varrho$$

ϱ = angle of internal friction

The active Rankine´s limiting stress ratio

$$\lambda = \tan^2(45° - \varrho/2)$$

introduced by Koenen rendered too small a value for the ratio p_h/p_v resulting in very high vertical pressures (Fig. 29). For the horizontal and wall friction pressures just the shape of the curves is wrong, whereas the endvalues agree with the measured ones. An exact method to determine the values of $\tan\delta$ could not be clarified, because the ratio p_w/p_h deviated about -5 % to + 47 % from the value determined by a shear apparatus (228). For practical purposes therefore $\tan\delta$ is assumed to be independent of wall friction and is a fraction of the angle of internal friction.

For the change in stress during discharge it can be stated in general that the vertical pressure declines, whereas the horizontal and wall friction pressures in the upper part of the silo rise suddenly and in the lower part slowly to a pressure maximum approaching a boundary value. Herewith $\lambda = p_h/p_v$ will be quite large in comparison with the filling and does not remain constant over the height of the silo, whereas $\tan\delta = p_w/p_h$ will be smaller but remains constant. The specification of a pressure function for the discharge is thus meaningfull only as an envelope of the maximum values (Fig. 30). This function can hardly be expressed mathematically. The

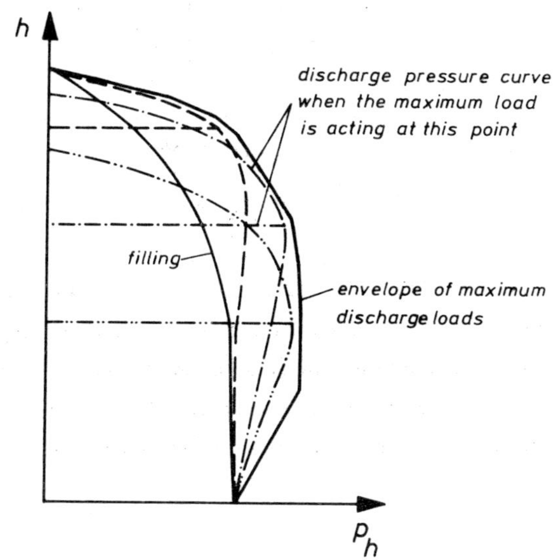

Fig. 30

reason why an e-function is specified in the standards is only to establish some standardized constructional formulas for filling and discharging. The inclusion of the coefficients λ and *tan* δ in the equation which can be estimated from the experiments is of descriptive nature and has no physical significance (229). The less significant change in the character of the curves shows a comparison between the really measured values and those calculated from the e- function Fig. 31.

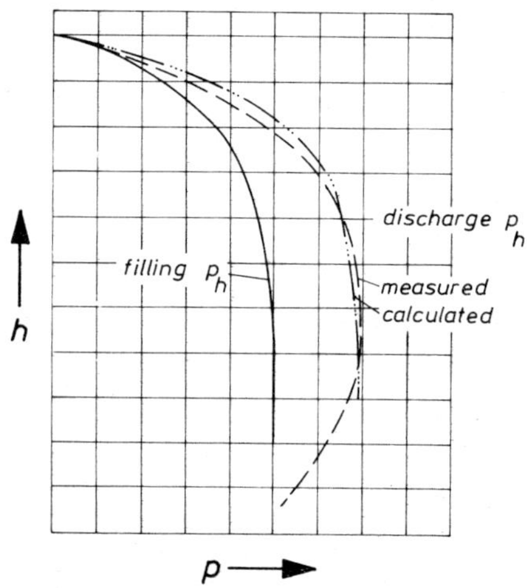

Fig. 31

For the calculation of the silo walls, the absolute pressure maxima are to be considered. These are given for the bottom loads during the filling and for the wall loads during the discharge of the bunker. For this case the German Standards provide the following coefficients, which are to be substituted in the equations of J a n s s e n and K o e n e n.

Table 5	δ		λ
	grained material av. grain diam. > 0,2 mm	powder material av. grain diam. < 0,006 mm	
filling	$0.75\,\varrho$	$1.00\,\varrho$	0.5
discharge	$0.60\,\varrho$	$1.00\,\varrho$	1.0

ϱ is determined by the usual methods of soil mechanics. Values of ϱ and $\tan\delta$ for various materials can be derived from the appendix of the book by Pieper and Wenzel (228). With $\lambda = 1$ for the discharge the hydrostatic stress condition is assumed, which refers to the ratio of maximum possible pressures during the bunker operation, and not to the actual stresses during the discharge.

In their book "Silos" M. and A. Reimbert (240) have derived equations for the static filling pressures. Although they are dependent upon the λ-values and are too low for $\lambda = tg^2\,(45^\circ - \varrho/2)$ due to the hyperbolic nature of the function, they lead to smaller vertical pressures and larger horizontal and wall friction pressures than in the formula of Janssen and Koenen (228). This book is specially valuable for the calculations of stress and size of wall and hopper for different forms of bunkers and for different materials of construction. The recommendations for the actual construction are subject to structural statics and cannot be covered here in detail. For this the elaborated list of references in Pieper and Wenzel (228) and Jenike and Johanson (135) can also be consulted.

The increase of discharge pressures are taken into account by Reimbert through overpressure coefficients, although the filling pressure remains constant when the bunker content is discharged from the top downwards according to the principle of last-in/first-out. Fig. 32 (240) shows a possible solution by installation of a central pipe with openings (left).

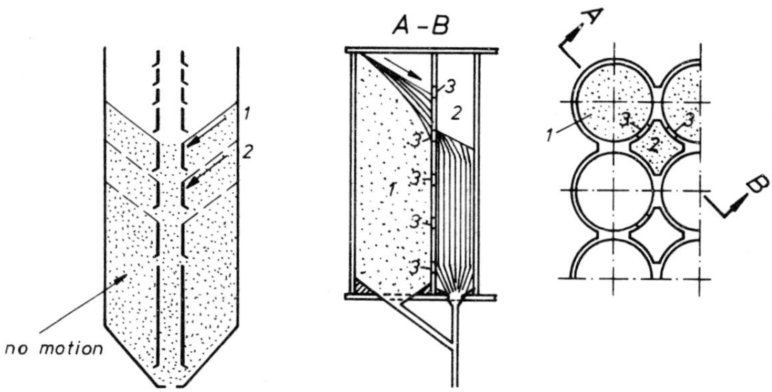

Fig. 32

The standards in the USSR (9) demand a reduction of filling loads. It is suggested to discharge the silo compartments through the radial intercells (center, right). Because the pressures, according to J a n s s e n and K o e n e n , depend linearly upon the hydraulic diameter F/U of the bunker cross-section, the wall pressures in the intercells are expected to be comparitively smaller than in the main silos. The standards used in the USSR are also based on the formula of J a n s s e n and K o e n e n with the function of R a n k i n e . The load increments are taken into account with correction factors. Herewith the effective load to be considered in the calculation of silo design becomes

$$p = p' \cdot a \cdot n / m$$

where

p' = the value calculated from the equation of J a n s s e n and K o e n e n ,

a = correction factor comprehending the load increments during the discharge, by temperature differences or by the collapse of arches, n = a not exactly defined "over-load factor" with a value of 1.3, which can be taken as a general safety factor,

m = factor considering the bunker construction.

Values of a and m for various bulk solids are listed in the appendix of the Russian Standard. They are graded according to the height of silos, in order to save building material. For cement it comes to (191)

for the lower 2/3 of the silo height

a = 2.0 m, m = 0.85

for 2/3 to 5/6 of the silo height

a = 1.0 m, m = 0.5

for the upper 1/6 of the silo height

a = 1.0 m, m = 1.0

3.2. Limitations of the Standards

Further investigations at the University of Braunschweig made it necessary to limit the validity of the German Standards (8), (191), (192), (227), (229), (230). They are valid only

a) for constructions, which allow a continuous flow without building up of any stable or dynamic arches. Otherwise it must be calculated with a linear pressure increment of $\gamma \cdot h$,

b) for a central discharge. The maximum wall pressure increments have been observed in zones, where the material flow takes place at the wall. Supplementary series of tests with eccentric discharge show an irregular character of maximum wall pressure near the outlet (Fig. 33). Especially the pressure peaks and the intense pressure drop in the lower part of the silo confirm the "pressure switch", theoretically derived from

Nanninga (210), Walker (308), (309), Jenike and Johanson (135) et al.

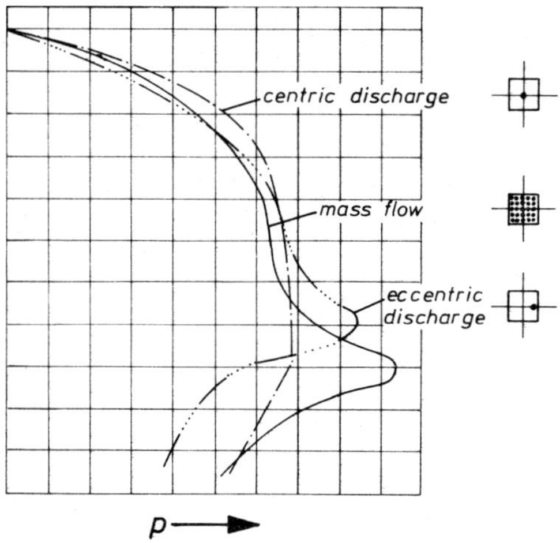

Fig. 33

These pressure peaks are distributed in a conventional core flow bin through the motionless zones on the wall.

A special case is represented by the bedding of very fine grained material in a bunker. The partial fluidization during the filling can lead to hydrostatic stress condition with absolute pressures, which lie remarkably above the standard values (191), (192). Through a gradual ventilation the material settles and the loads lower to the standard values. The wall pressures measured immediately after the filling exceed the maximum expected loads, according to the standards, by about 3.13 times. It is apparent that the overpressures during filling are essentially determined by the filling rates. Martens (191) has found that for every bulk material a characteristic limiting filling rate v_o exists, below which no pressure increase takes place. He found the following values for the materials tested by him:

limestone powder	v_o = 1.4 m/h
cement	v_o = 2.6 m/h
wheat powder	v_o = 4.8 m/h

The hydrostatic bedding pressures can be estimated roughly according to the formula:

$$p = 0.8 \cdot \gamma \cdot z$$

z is the silo depth measured from the surface of the bulk material, at which the overloads are reduced by consolidation of the bulk material. The values of z can be calculated by the formula

$$z = (v - v_o) \cdot t$$

with v = actual filling rate and t = 1 hour.

For these powdered materials special discharge techniques are developed (see Chapter 6. 2.), which provide for equipment to blow air in the lower part of the silo. The flow of the bulk material will be improved through the controlled fluidization in bottom region i. e. the bulk material will approach the behaviour of fluids. The height of fluidization zone h_f, in which extra hydrostatic pressures appear, can be evaluated according to Martens as follows:

$$h_f = 1.33 \cdot p_u / \gamma$$

p_u is the blowing pressure just above the bottom of the silo, which at the same time represents the full load for this region and which can be determined by the compressor pressure less the pressure losses within the pipes and entrance into the bottom of the silo.

$$p_u = p_c - (\Delta p_p + \Delta p_b)$$

In the fluidized zones the overpressure is practically linearly reduced. Fig. 34 (191) shows the pressures in closed (left) and open (right) bunkers in comparison with the corresponding standard values.

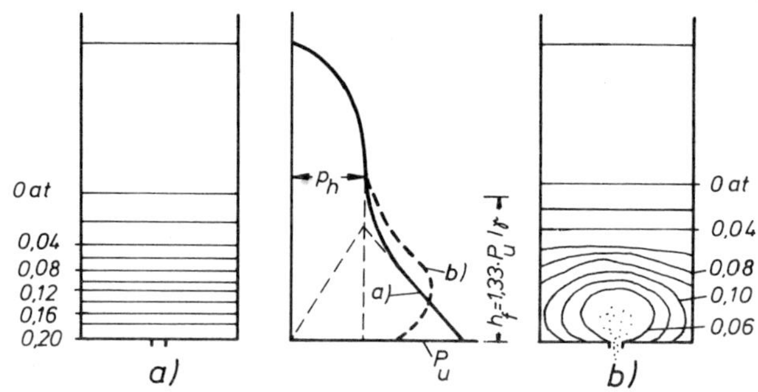

Fig. 34

As a basis for calculation Martens suggests that the standard pressure values are taken for the area between the surface and the height h_f and the pressure p_u for the bottom region. For the fluidized zone it suffices to interpolate linearly between the bottom (h = 0) and h_f.

3. 3. Pressures in Mass Flow Bins

The definition of mass flow states, that the movement of bulk materials takes place along the wall over the whole height of the bin. For these types of bunkers, the calculation standards to calculate the silo loads, as discussed in Chapter 3. 1., are insufficient. It was already mentioned in Chapter 3. 2., that a pressure peak occurs, which lies considerably above the pressures calculated according to the standards. Because of this the construction of an existing bunker has to be calculated again (Chapter 2. 1.), in case improvement measures lead to mass flow. Jenike and Johanson (135), in their fundamental study "Bin Loads", give design data, which they outline in two examples. They explain the phenomena of pressure peak as follows.

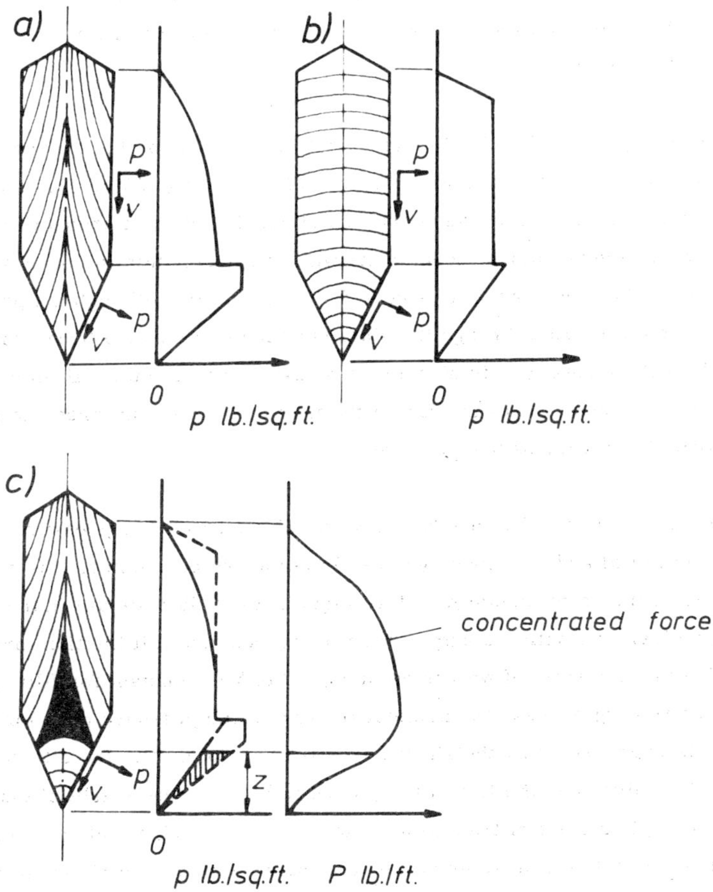

Fig. 35

During the bedding the active state of pressure exists, since the material contracts under the increasing loads. The major pressure lines are nearly vertical (Fig. 35 a) (135). During the flow the passive state of pressure sets in, while the material in the hopper expands in a vertical direction and contracts in horizontal direction. The principal pressure lines are practically horizontal or bridge over the outlet (Fig. 35 b). At the beginning of flow (Fig. 35 c) a reorientation from the vertical to the horizontal direction occurs at the outlet.

The pressure peak, called by Jenike and Johanson as the "concentrated force", is caused by the black colored volume in Fig. 35 c. This volume is in a state of transition between active and passive state of pressure and is diverted to a very narrow zone on the wall of the bin. Similar processes take place in the flow channel of a core flow type of bin, although the pressure peak will be distributed through the stationary material over a broad zone of the wall. Instrumentally it is difficult to measure the absolute values of these pressure peaks, because during the flow they travel upwards and are concentrated in a very narrow wall zone.

The absolute value results theoretically from Fig. 35 c. The pressure curves for initial and flow loadings are plotted above each other. The area below each curve represents the total weight of the bulk material in the bin. It must be identical with the sum of the wall forces and wall friction forces. Assuming that the ratio of wall friction $v/p = \tan \phi'$ is constant for filling as well as flowing phases, the numerical value of the pressure peak for the height z is represented by the shaded area in Fig. 35 c. It is the difference between the initial and the flow loadings and is equal to the weight of that volume being in a state of transition. The curve on the right side of Fig. 35 c shows the qualitative character of the concentrated force, which has to be superimposed over the envelope of the curves for the initial and flow pressures, in order to determine the effectively occuring wall loadings.

The knowledge of maximum possible wall loading is especially important. To give an impression of the magnitude up to which the maximum pressure peak p_{max} can exceed the active pressure p_a just above the pressure switch, Fig. 36 (135) shows p_{max}/p_a as the function of the angle of wall friction.

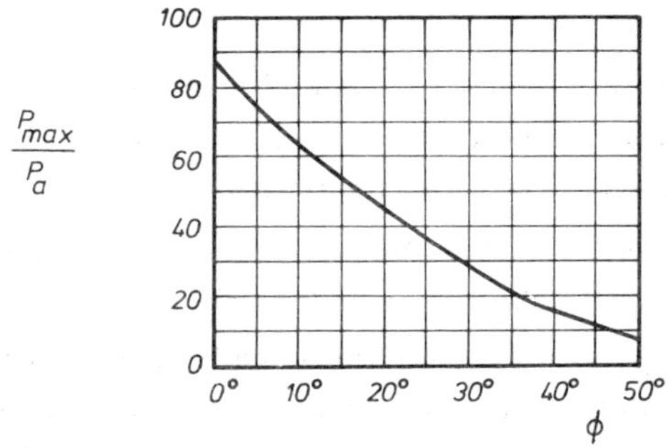

Fig. 36

The values of p_{max}/p_a for the effective angle of internal friction $\delta = 50^o$ are calculated by J e n i k e and J o h a n s o n from the equation:

$$\frac{p_{max}}{p_a} = 2 \tan\delta \, \frac{1 + \sin\delta \cdot \cos(\varrho + \Phi')}{1 - \sin\delta \cdot \cos(\varrho - \Phi')} \cdot e^{(\pi/2 - \varrho)}$$

where $\sin\varrho = \sin\Phi' / \sin\delta$

(The symbols are the same as in Chapter 2. 1.)

A function of p_a is not given directly. Although according to size, it can be approximated from the standard specifications for silo loads during filling the bunker mentioned in Chapter 3. 1., which have shown quite good conformity between values calculated from formulas and experimental values (228). Because the region of the bunker in the transition between cylindrical and conical part is the most critical zone, the maximum value of the filling load curve can be assumed to be p_a . Of course it must be reduced to the circumference of the bin, in order to achieve the same dimension (pounds per foot of circumference) as the concentrated force p_{max}.

Jenike and Johanson have further shown how to calculate the position and value of the pressure maxima for the initial loading, flow loading and the concentrated force. A more detailed discussion of this would be too extensive for the scope of this chapter. A new and detailed publication, to be published as a report of the American Iron and Steel Institute, is announced by Jenike and Johanson for some later date. The publication will include detailed description of the theory of silo loads as well as a complete set of charts for bins with and without surcharge.

4. Shapes of Bins and Bunkers

Bins, bunkers and silos are used worldwide as receptacles for the storage of bulk solids, but the terms are not clearly defined in all countries. Usually the terms bins and bunkers mean the same for vertical and horizontal containers, with or without hoppers, and are mainly used in certain branches of the industry, such as the heavy industry (height $\approx$ width). The term silo is used for containers such as in grain and cement storage, which are narrow and high (height $\gg$ width) (10).

In considering the shapes of bins and bunkers, it is essential that the design of the bin must be closely related to the flow properties of bulk materials in order to get a mass flow. Based on the flow properties, Kvapil (166) has divided bulk materials into four main groups:

A = free flowing bulk material,

B = bulk material with inferior flow characteristics,

C = lumpy bulk material,

D = material with very poor flow properties.

Depending on this classification he proposes various shapes of bunkers with different ratios of bunker diameter (d) to outlet diameter (e) (Fig. 37) (169), (302).

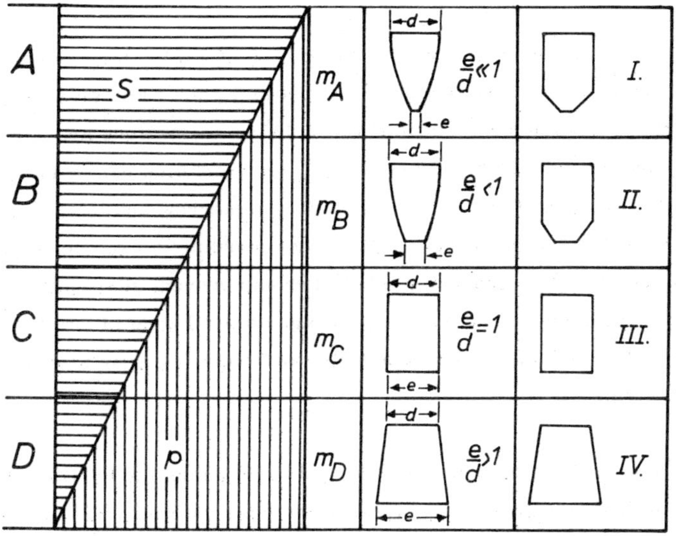

Fig. 37

Figure 37 also indicates the characteristics of the flowability of the granular materials classified by A, B, C, D which is

$$m_A > m_B > m_C > m_D$$

According to this illustration, the regions of primary (p) and secondary (s) motions, which were defined earlier in Chapter 1.1., are increasing or decreasing. That means, that a free flowing bulk material A, which exhibits primarily a secondary motion, can be stored in hoppers with only a reasonably small outlet. In case of D, due to the increase of the primary motion, the conditions are reversed. For a more detailed classification see Chapter 1.3.

4.1. Capacity of Storage

One of the most important factors for the choice of a silo, bin or bunker is the storage capacity, required by the industrial plant. For small capacities single receptacles are sufficient. Their cross-sections are mostly round, square or rectangular. Especially cylindrical bins can be built according to the method of Jenike resulting in narrow and tall structures with hoppers of certain wall inclinations, or according to experience with hoppers steep enough to guarantee mass flow and prevent segregation. In cases where the hoppers are not steep enough, a dead zone is created around the outlet reducing the flow activity and requiring special devices for discharging.

For greater capacities bins can be enlarged by widening the horizontal dimension. In this case several drawpoints are possible which can be opened simultaneously or one at the time. Fig. 38 and 39 show the movement of the material and the intersection of the primary ellipsoids in a model bin. Fig. 38: Consecutive opening of the outlets, Fig. 39: Simultaneous discharge from all openings (163).

Silos consisting of a group of vertical storage bins are well known. They can be round, square, rectangular or hexagonal and may be individual units of clusters with or without interstices (Fig. 40) (117), (273. In groups like

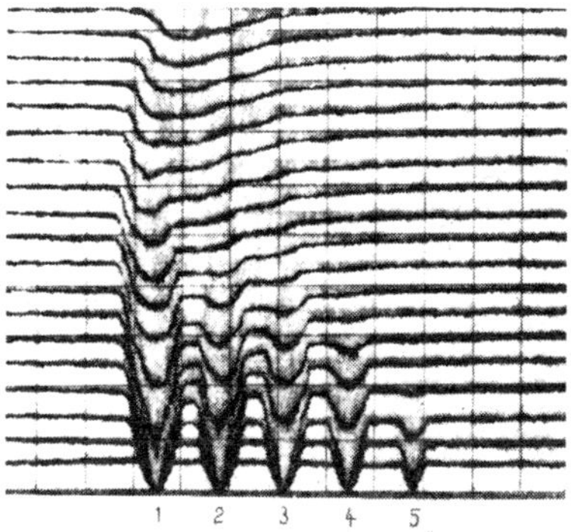

Fig. 38

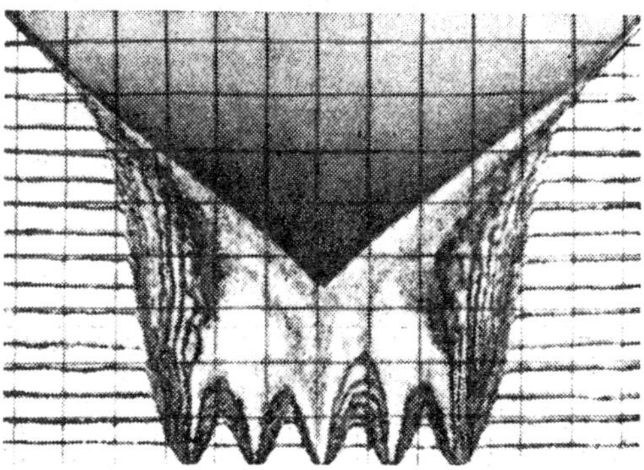

Fig. 39

this, they guarantee a large storage capacity. With an automatic handling system, a silo requires only a few men for operation. A greater variety of materials (for example: types of grain) can be stored since the silo as a whole is divided into a large number of compartments.

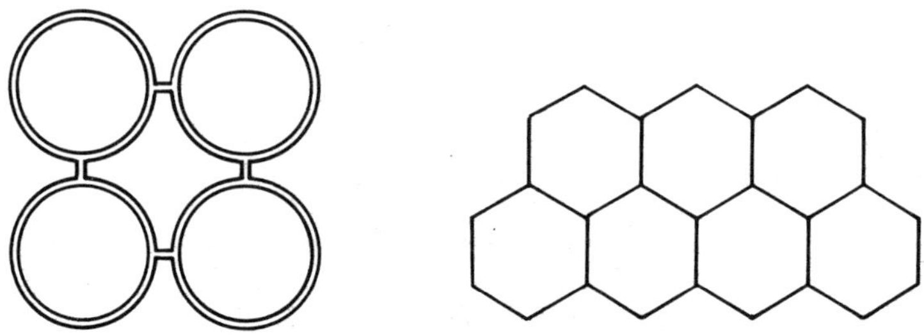

Fig. 40

A similar arrangement with larger bins enables homogenization of the bulk material by discharging a certain amount in a certain sequence. Fig. 41 shows a sketch of a compartmented bin for continuous blending of raw coal according to a proposal by H i r s t (121). Another type of a homogenizing bin is designed as so-called "organpipe type bin" for space saving purposes (Fig. 42) (144), (282) (More details on homogenization see Chapter 7.1.).

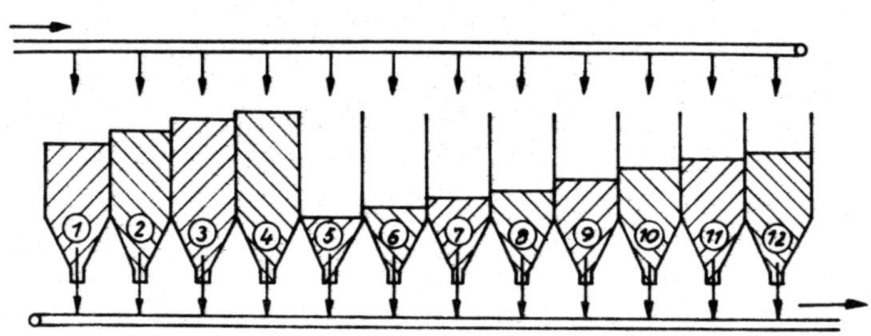

Fig. 41

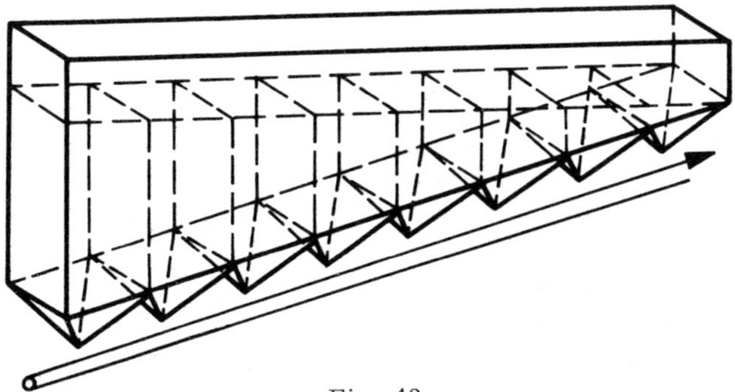

Fig. 42

Flat-bottom bins have the advantage of requiring less height for a certain volume, and less space for their construction. Since the stored material does not flow out completely, the discharge must be assisted by mechanical or pneumatic means. Bins of this type vary considerably in shapes, dimensions and dischargers: round, square, rectangular, slot bunkers, horizontal bunkers etc. They are used for example

a) for materials with very poor flow properties, because even for other shapes a discharger would be necessary,
b) for the storage of fine-grained materials or powders which are to be fluidized for extraction,
c) in case of structural conditions which do not allow high containers or where a hopper would increase costs considerably.

Concerning the dischargers more details are given in Chapter 6, together with the function of these devices.

Inclined bins or silos are occasionally used in order to reduce degradation of the stored material. Fig. 43 (294) shows the flow pattern in a model with

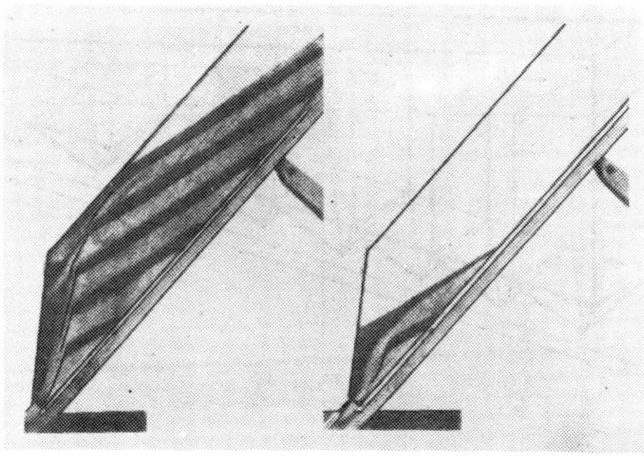

Fig. 43

different layers during flow and emptying. There is a core flow along the left wall allowing only the top layers to be discharged. This operation causes a considerable segregation. Several flow patterns (163), (263) show this behaviour as a function of the inclination of the bin with special consideration of the development of ellipsoide of motion.

Economic considerations in respect to hopper design, withdrawal feeders, and flow-aid devices have been outlined by Vitunac (306).

4.2. Determination of Volume

Since bins, bunkers and silos are involved in technical processes, it is important to determine their correct volume. They must be large enough to store as much as necessary, but, on the other hand, over-sized bins are not economic because of their construction and maintenance costs.

The diagram in Fig. 44 (279) gives an example for finding the correct bunker capacity in relation to time. Suppose a bunker is fed, during a shift of eight hours, every two hours with a charge of 100 tons (step line 0 - A) and is discharged from one o´clock continuously (line B - C), the largest distance between these lines mark the minimum content of the bunker with 150 tons. The ordinates of the shaded area show the content during the whole shift.

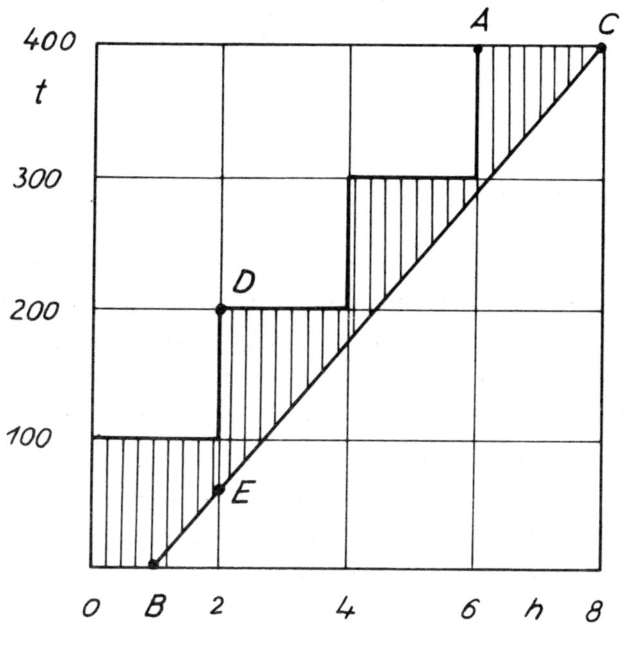

Fig. 44

Further investigations in this field based on operations research are limited. Elbrönd (87) suggested the use of the frequency of interruptions when feeding and discharging the bunker, the volumetric efficiency in relation to feeding and discharging per unit time and the over-all efficiency of the storage system. Ziems (331) discusses the dimensioning of transport and storage units for transloading points. Other authors (49), (283) have derived relationships regarding the utilization of belt conveyors with intermediate bunkers as a function of time at a given bunker capacity and a known trouble pattern of the conveyors. From these relationships optimum bunker capacities may be determined. Stoyan (284) describes a calculation method for the optimum size of bins in systems containing one bin, considering that the accidentally varying feed and discharge are stationary probability processes, i. e. that they have constant mean values. Russian scientists (259) calculated volumes of bunkers for underground mines for homogenization at loading points near coal faces, applying the queuing theory.

4. 3. Cross Section of Bins

Usually a vertical bunker consists of an upper part with a constant cross section which is also called "bin" and a lower part, the so-called "hopper". The most common cross sectional shapes of bins are square, rectangular and round. Center, side, corner and slot discharge are some of the hopper types used in bin constructions. The design of bin configuration should be based on the flow properties of the materials to be handled, if possible using the method of Jenike (133). Other styles of hoppers include multiple openings at the discharge, bifurcated dischargers, double hoppers and hyperbolic outlets mentioned also in other chapters.

Research work has proved that the shape of the cross section of a bin has no influence on the discharge rate (41), (114); it only influences the pattern of outflow. In case of a circular or elliptical cross section, the layers of the stored material move downward horizontally if the bin is constructed for mass flow. Fig. 45 (243) shows straightlined moving layers in a cylindrical model, after having removed the front half. If the cross section is square

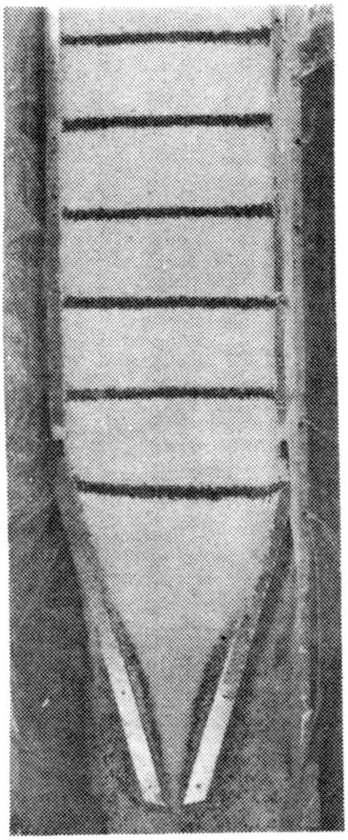

Fig. 45

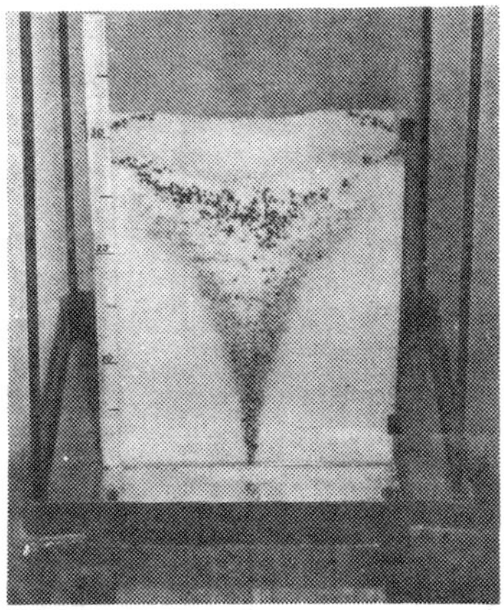

Fig. 46

or rectangular, a slower motion can be noticed in the corners so that the mass flow pattern is reduced. This disadvantage is even more serious if the bottom of the bin is flat, i. e. the bin has no hopper (Fig. 46).

4.4. Shape of Outlets and Hoppers

4.4.1. Outlets in Flat-Bottom Bins

Although the flow pattern of bulk material is discussed in other chapters, some discussion of that subject must be made here because of its influence on the shape of the outlet. In experiments with a flat-bottom bin model and a central outlet, the movement of layers could be followed by motion pictures. The results are shown in Fig. 47 (243). As long as the diameter of the outlet is small (left) there is an irregular flow, referred to as core flow.

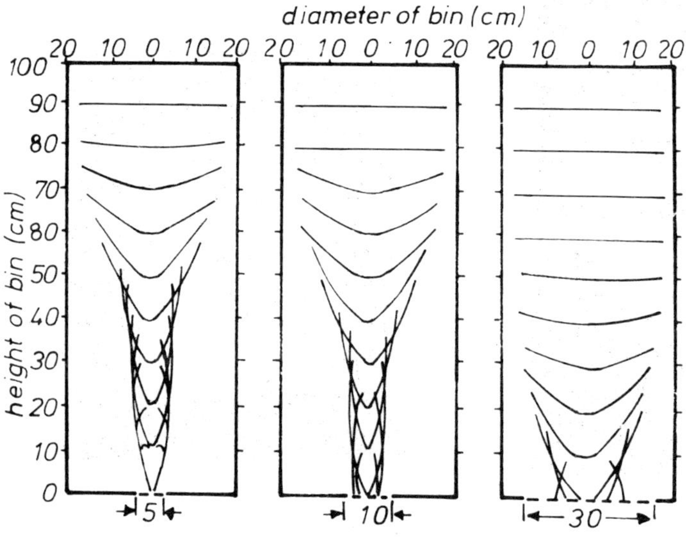

Fig. 47

The larger the outlet becomes, the better the flow conditions are. This means that there is a more regular motion in the whole bin, approaching mass flow, which is the preferred type of flow.

Since the diameter of the outlet affects the flow pattern of a material in a bin, tests were run to determine the maximum radius $\overline{Y}$ of the boundary between flowing and nonflowing material for flat-bottom bins in relation to the diameter D of the circular discharge opening (Fig. 48) (104). Provided the material is free flowing, the width of the flow pattern could be predicted by the equation:

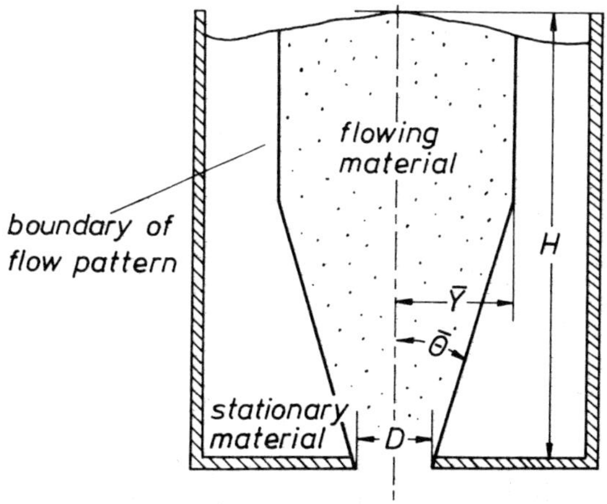

Fig. 48

$$2\bar{Y} = D + 2\tan\bar{\Theta}\left(\frac{H - A\frac{D}{2}}{1 + A\tan\bar{\Theta}}\right)$$

when H > AD/2 and

$$2\bar{Y} = D \quad \text{when} \quad H \leq AD/2$$

The result is a function of the opening size (D), the angle of flow pattern boundary with vertical axis ($\bar{\Theta}$), height of material in the bin (H) and a derived dimensionless flat-bottom bin factor (A).

Another observation deals with the following phenomenon (193): When a free-flowing material is flowing through an outlet with the diameter D in the centre of a flat-bottom bin, the flow range is decreased by a "statistically empty space" k adjacent to the inner edges of the aperture (see Chapter 2.2.2.). Brown (59) has found a numerical coefficient in the flow relation which is a function of an angle of sliding associated with flowing powders. ϕ is the "angle of elevation of the boundary separating the central flowing zone from the surrounding stationary material, and is greater than the drained angle of repose" (193).

It has been experimentally determined that the discharge rate for a flat-bottom bin is greatest when the aperture is circular (114). Fig. 49 shows results of some tests with apertures of various shapes but with the same area, such as rectangle, ellipse, square, circle. The discharge rate is a

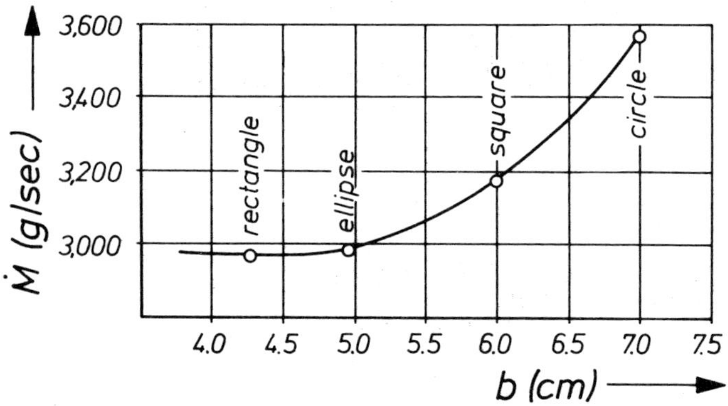

Fig. 49

function of the smallest dimension of the outlet. The smallest dimensions are plotted on the axis of abscissae. The discharge rate increases as the smallest dimension of the outlet increases. - Similar tests were also run by Runo (257).

4.4.2. Hoppers

The design of a hopper is very important for the success or failure of the bin function. A hopper changes the direction of flow of particles in a bin.

The material, which was formerly spread over the entire cross section at the upper part of the bin, must converge and flow through a much smaller outlet of the hopper. The ability of the material flow to converge towards the outlet is largely determined by the construction and shape of the hopper.

The upper part of the hopper coincides generally with the shape of the bin cross section. Various examples of hopper shapes are shown in Fig. 50 (272). The most common shape is a symmetrical, conical hopper, used in

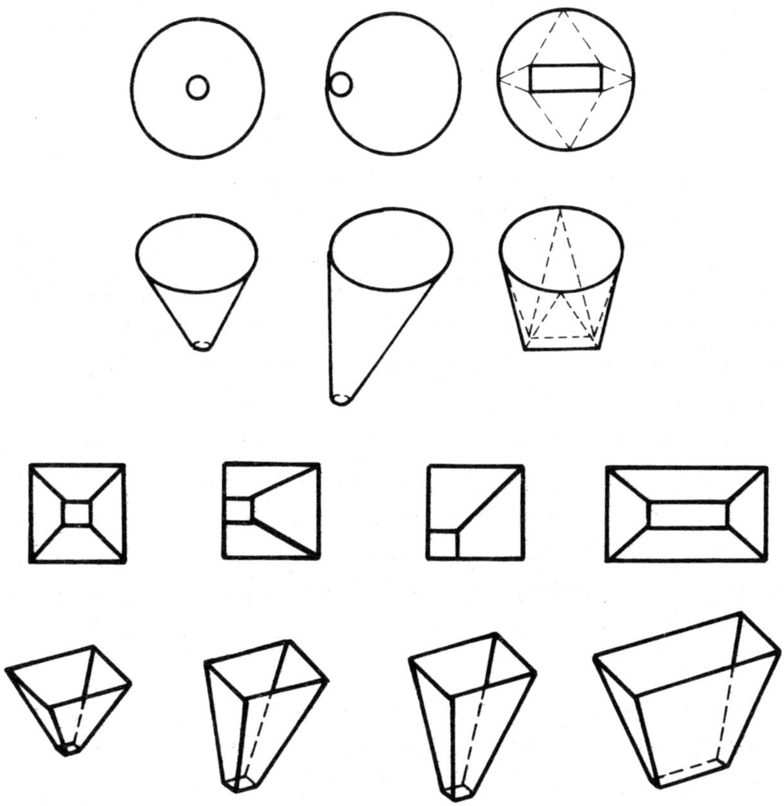

Fig. 50

connection with round bins. They are relatively easy to construct and perform satisfactorily with fairly free flowing materials. If the hopper slopes are not steep enough, funnel flow takes place on a last-in/first-out basis. As mentioned earlier, Jenike has developed a method to calculate the minimum diameter of a hopper outlet to prevent flow stoppage (74), (130), (132). If the diameter of the outlet is increased, the hopper becomes more suitable for movement of less flowable materials, as greater material strength is required to sustain an arch over the larger opening.

More detailed information, based on a literature study, has been presented in a paper by Bates (32).

An eccentric hopper discharge is used frequently for round, square and rectangular bins when difficult flowing materials are stored. As one side of the hopper walls is vertical and the outlet is adjacent to it, the occurrence of stable arches is less frequent because the material tends to slide down on the vertical wall. Observation of hoppers with side discharge show that provision must be made to prevent buckling of the wall directly above the discharge opening, because the load on the vertical wall is higher than on the opposite side where the bulk solids are at rest (28). When a bin with free flowing material is emptied through a vertical opening in the wall, the discharge rate is nearly as great as through a horizontal opening of the same size in the bottom (28). The grain flows at a constant rate during discharge of the bin. Consequently, the rate is independent of the height of the material in the bin.

In connection with this phenomenon it should be mentioned here that research work on the rate of discharge as a function of the height of the bunkered material has shown different results. If the height of the material exceeds four times the diameter of the bin, the rate of discharge is constant. Small variations are explained by the density fluctuations of the stored material. However, considerable variations in discharge rate occur when the level is low. S c h w e d e s (267) makes reference to respective experimental work and publications.

Round, square or rectangular hoppers with a slotted outlet are sometimes used to prevent funneling, by making the outlet dimension (length or diagonal) almost equal to the bin diameter. For this purpose J e n i k e (132) has proposed a chisel-shaped hopper (Fig. 51). To remove the material out of this hopper, a feeder is necessary over the whole length of the slot.

If square or rectangular bins with hoppers of the same shapes are used, the corners and valleys in the structure will cause a hang-up of material,

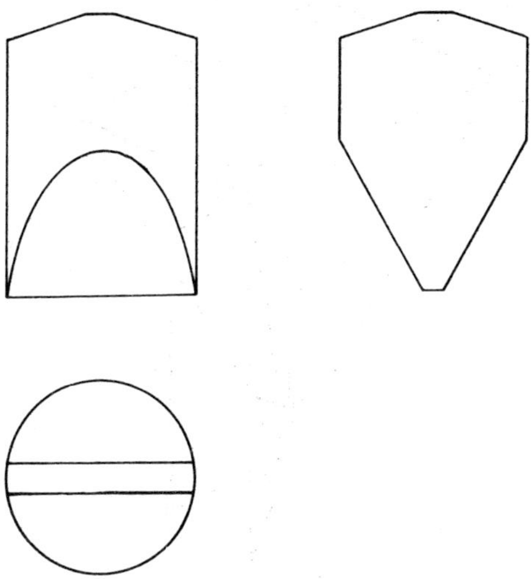

Fig. 51

impeding flow. It is important to place fillets in these corner areas so that obstructions can be minimized or eliminated (74).

As the slope of a hopper wall is of significant influence on the flow pattern, the literature suggests many recommendations of making the hopper slope at least as steep as the angle of repose of the stored material (178), (179), (214), (216), (256) or better about 15 degrees steeper (41), (156), (241), (291). Theoretically the minimum hopper slope should correspond with the angle of wall friction. This will only guarantee that the hopper empties completely, but it will not develop a condition suitable for mass flow.

In case of rectangular or square bins with pyramidal hoppers, the bin valley angle ϕ (Fig. 52) (173) influences the flow pattern,too. Whilst the

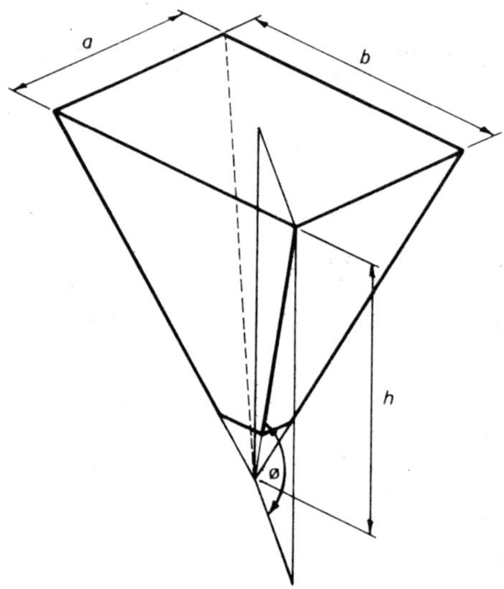

Fig. 52

sloping sides may appear to be steep enough, the bin valley angle may present a problem and retard flow. For a rectangular bin the bin valley angle is given by

$$\phi = \tan^{-1}\left[\frac{2h}{\sqrt{a^2 + b^2}}\right]$$

and for a square bin, when a = b, by

$$\phi = \tan^{-1}\left[\frac{h\sqrt{2}}{a}\right]$$

h = vertical depth of hopper (ignoring outlet cut-off)

b = the longer side

a = the shorter side

ϕ = the bin valley angle measured from the horizontal

On the basis of many experiments, the discharge rate is proven to change when the hopper slope becomes steeper.

Fig. 53 shows the flow rate as a function of hopper slope for different bulk materials. In the lower range of hopper slope angles the discharge rate was found to be irregular (243). Laforge and Boruff (172) explain this phenomenon by the fact that the particles which move relatively slow from the outer region into the vertical column of discharging material restrain the velocity of efflux. In Fig. 54 the paths are outlined by the points A and B. This figure also shows the more regular motion of the particles with increased hopper slope. Some other examples showing the influence of hopper shape are given in Fig. 55. These experiments and those shown in Fig. 56 were run in connection with the testing of the movement of crowds of people through narrow exits which could be compared in simplified form with hopper problems (80), (218). Tests with sloping bottoms and central outlets have shown, that at an angle of 60 degrees (d) the time of efflux is cut 60 % of

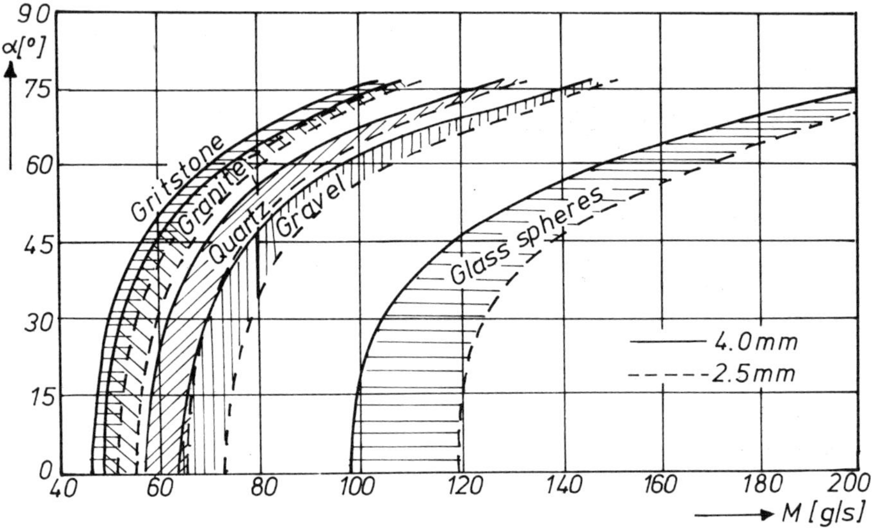
90
α[°]
75
60
45
30
15
0
40
60
80
100
120
140
160
180
200
M [g/s]
Gritstone
Granite
Quartz
Gravel
Glass spheres
4.0mm
2.5mm

Fig. 53

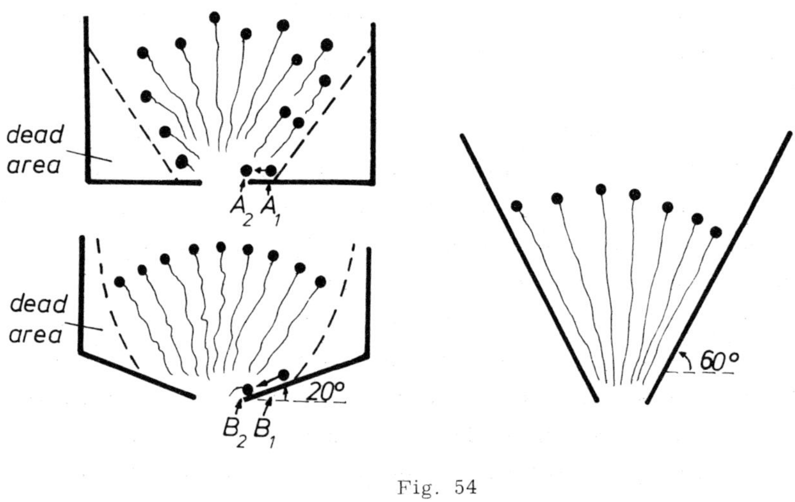
dead
area
A2
A1
dead
area
20°
B2
B1
60°

Fig. 54

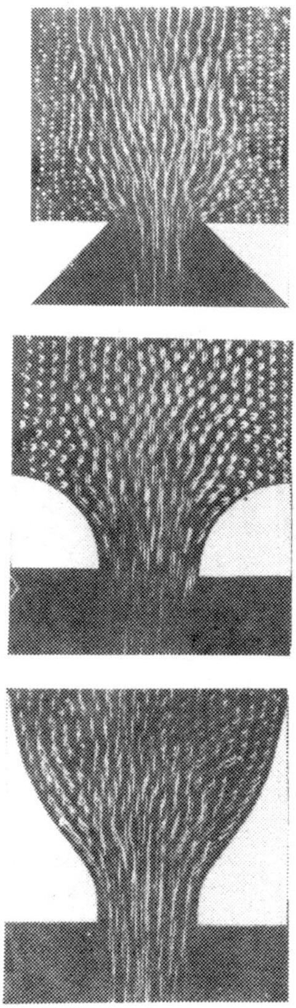

Fig. 55

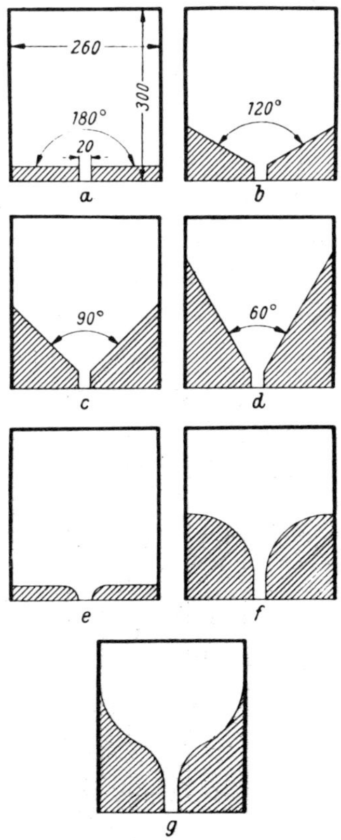

Fig. 56

that needed for a flat bottom (a). The funnel-type outlet (g) reduces the time to about 30 %. Even a very slight rounding of the outlet (e) improves the efflux and consequently the discharge rate. The photographs of the movement of spheres in different models (Fig. 55) also show the advantage of rounded or funneled outlets.

Richmond (251), (252) defines the "best hopper profile as the one which can transfer a given bulk material by gravity flow from some given entrance opening to some given smaller exit opening in the least vertical distance". Based on this definition, he derives equations for plane (rectangular) and for axially symmetric (circular) hoppers allowing computation of the hopper profile as a function of the mechanical properties of the bulk material and the material-wall interface friction. The above mentioned references explain the method of calculation in details. In this case, the "best" hopper is concave with an exponential curved profile. An experimental verification was obtained for four different profiles as shown in Fig. 57. The criterion of

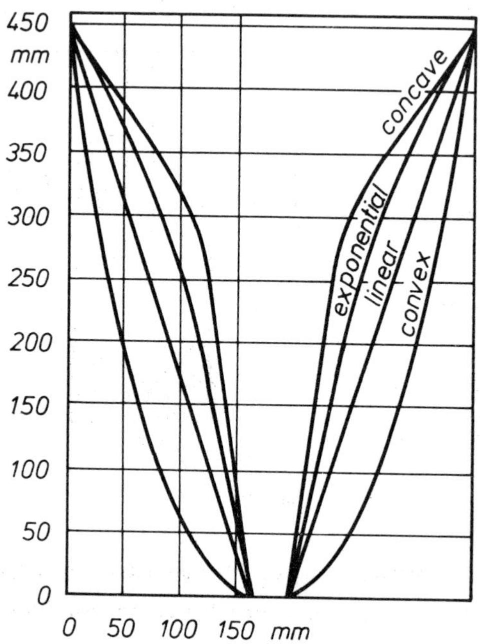

Fig. 57

evaluation was the width of the outlet which allowed the material to flow out freely. The exponential profile permitted flow at a slot width of about one third of the width needed for the convex curve and one half of the width for the linear and concave profiled hopper. - An exponential profile is also proposed by P e s c h l (225).

L e e (182), (183) has published some results of his theoretical investigations on the best shapes for gravity-type coal hoppers. He has found a hyperbolic profile to be most suitable which agrees with the general concave shape of Richmond. More mathematical examples for the design of a hopper with hyperbolic walls for fine grained coal and the interaction with the pressures by the bulk materials are given in (332), (333), where the authors based their work on the theory by L e e . They are too voluminous to be published here.

An important criterion of L e e´s theory to prevent arching is the relative change of cross-section which should be nearly constant towards the outlet or much smaller than required in the upper zones of the hopper.

Based on this relative change of cross-section, K n e s c h k e (154) proposes for a rectangular bunker with a very fine grained material, that the longer sides have hyperbolic and the shorter sides have exponential profiles. This combination, which ensures a considerable increase of bunker activity, was developed because it was found that a hyperbolic profile of the shorter sides does not improve the flow as much as an exponential one.

An example for profiling a hopper under a traveling-grate sintering machine is given by J o h a n s o n and C o l i j n (137). The sub-atmospheric pressures which exist in the dust hoppers cause a countercurrent flow of air when the discharge valve is opened, thus retarding the movement of material. If the hopper cannot be made sufficiently steep because of space limitations, then a dust hopper made from concave panels, as illustrated in Fig. 58, could be of considerable advantage.

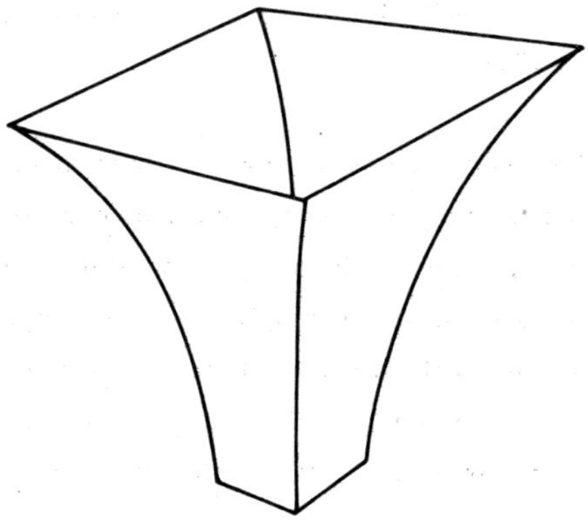

Fig. 58

The results of the research work by Gardner (100), (101) concerning improved hopper profiles can only be evaluated numerically and they are represented in diagrams.

The afore mentioned results of experimental and theoretical research work have been evaluated and veryfied by manufacturers (F 5)[x]. Among other devices connected with the storage of bulk materials, they construct hyperbolic outlets for all types of bins, tanks and silos which assure positive flow of difficult materials, granular and lumpy materials without the use of accessory equipment. Some more examples for specially designed bunkers for the storage of poorly flowing bulk materials like power plant coal and fertilizers are given by the following references (27), (139), (184), (186), (234), (238), (246), (327). These examples cannot be described in detail because they are too different.

----

[x] See list of manufacturers, pp. 277 - 280.

Regarding further constructions of discharge promoting hoppers, it is particularly important in rectangular and square bins that the hopper walls must be arranged in such a way that they always intersect with a vertical wall (186), (297). Examples in Fig. 59 are given by Theimer (298).

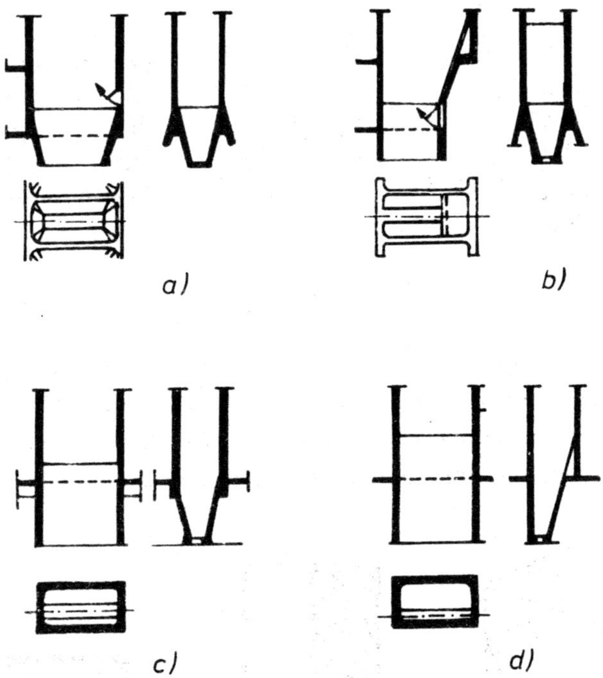

Fig. 59

A discharge hopper with four inclined panels (a) is not recommended because the bulk material builds up itself in the grooves between the panels thus causing funnel flow. The sketches b and c show the improved constructions with two or three sloping panels, while d shows a hopper with one sloping wall as the best method.

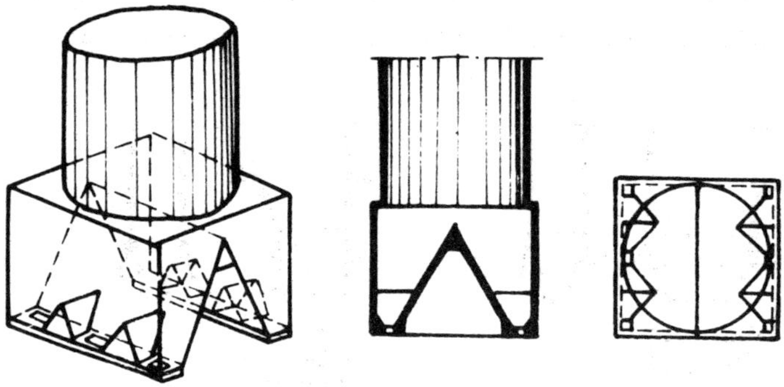

Fig. 60

Fig. 60 illustrates a specially designed discharge compartment for silos with six separate outlets. The combination of a cylindrical silo and a square hopper allows the enlargement of volume (F 42). A silo with a split-bottom hopper is shown in Fig. 61. By this construction the bulk material in one

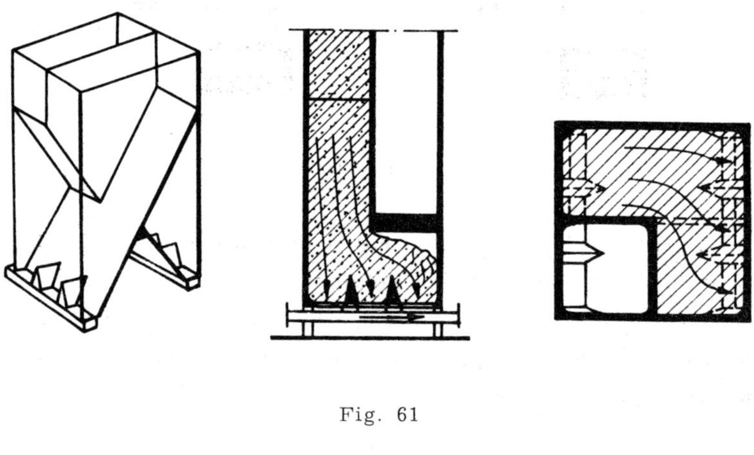

Fig. 61

rectangular silo compartment can flow in the region of outlet under the inclined hopper panel of the next silo (296) (F 3). Another construction has a special arrangement of four panels either for circular or square cross sections (Fig. 62). The bulk material moves towards the outlet step-wise and undergoes a slight cork screw motion which enables mass flow (F 32).

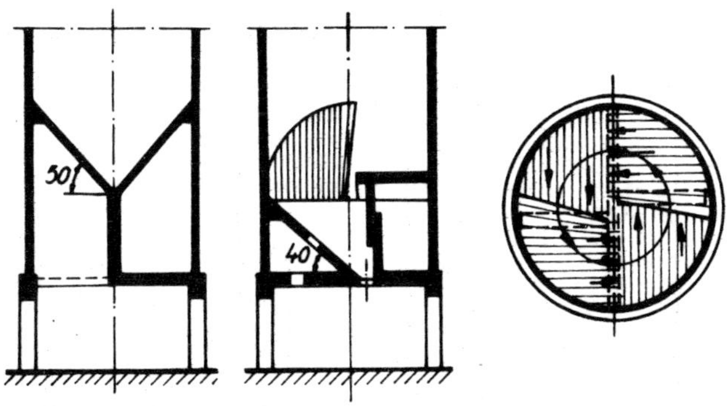

Fig. 62

4.4.3. Inserts

Most of the improperly designed bunkers have a low flow activity which may cause difficulties in continuous operation, such as segregation and passive zones where sometimes chemical alterations may occur. An improvement can be achieved by providing larger discharge openings. However, there are limits to the opening size because of the increasing pressures acting on the area of outlet and on the discharge devices. Conical inserts may improve the flow pattern in the whole bin, provided they are in a favourable position.

In his theoretical calculations J e n i k e found that plane flow symmetry promotes flow more than axialsymmetric flow (see charts Chapter 2.1.3.).

This improvement may be achieved by an insert in the hopper, since the ring-shaped area between the hopper wall and the insert can be assumed as a slot outlet.

Experimental and practical investigations by Kvapil and Tanaka (168) have shown that for free flowing bulk materials the use of horizontal baffles as inserts over the outlet can produce a larger active flow zone (Fig. 63).

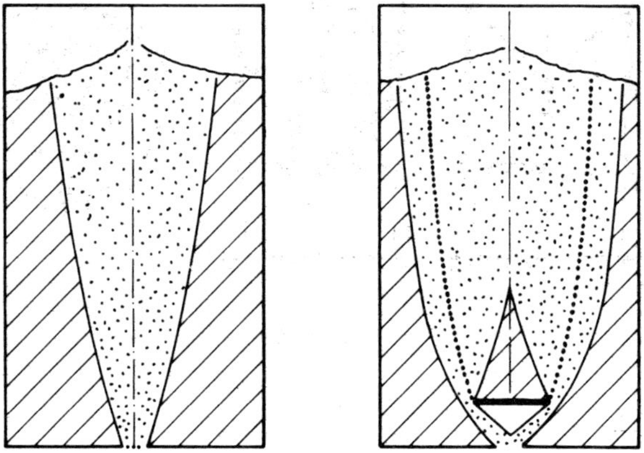

Fig. 63

Similar experiments have shown comparitive flow patterns with and without eccentric baffle plates in converging discharge channels in coal bunkers (168). A large insert placed near the transition of the bin and hopper may develop a mass flow pattern and an insert placed near the outlet may be useful to eliminate piping and arching of bulk solids.

A hollow space exists below the baffle. For a specific diameter of the outlet and angle of repose of the bulk material, a certain sized baffle may favorably influence the discharge operation if it is properly located over the opening.

The amount of bulk material acting on the outlet is only that volume between the outlet and the hollow space and the pressure is independant of the material height in the bin.

Kvapil and Tanaka (168) have calculated the minimum baffle dimensions to improve flow. Some important relationships are quoted in the following.

According to Fig. 64, the height of material above the closed outlet is

$$H_S = 0.5 \cdot d \cdot \tan\varphi$$

where d = diameter of the outlet

φ = angle of repose of the bulk material

and the thickness of the layer is

$$H_B = \frac{d}{2 \cdot \cos\varphi}$$

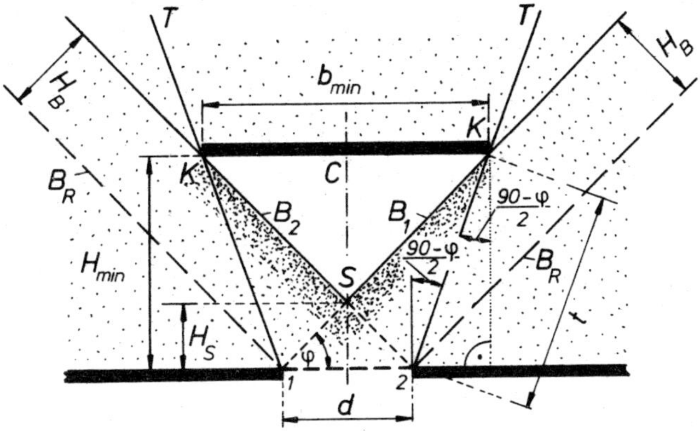

Fig. 64

For practical purposes, it may be assumed that the slopes of the material follow the slip lines T-1 and T-2 having an angle of 0,5 (90 - φ)^o with the vertical. Hence, the smallest diameter b_{min} of the baffle is defined by the point of intersection K of the lines T-1 and B_2, resp. T-2 and B_1:

$$b_{min} = d + 2 \cdot t \cdot \sin[0{,}5(90 - \varphi)^\circ]$$

The minimum height of the baffle over the outlet is

$$H_{min} = t \cdot \cos[0{,}5(90 - \varphi)^\circ]$$

where t is

$$t = \frac{H_B}{\sin[0{,}5(90 - \varphi)^\circ]}$$

Baffles which slope symmetrically towards the outlet will also force the material above to move downward in approximately parallel lines. Fig. 65 shows a sketch of a bin equipped with four inclined baffle plates and a central deflector (200). The slope of the baffles should be about 10 degrees steeper

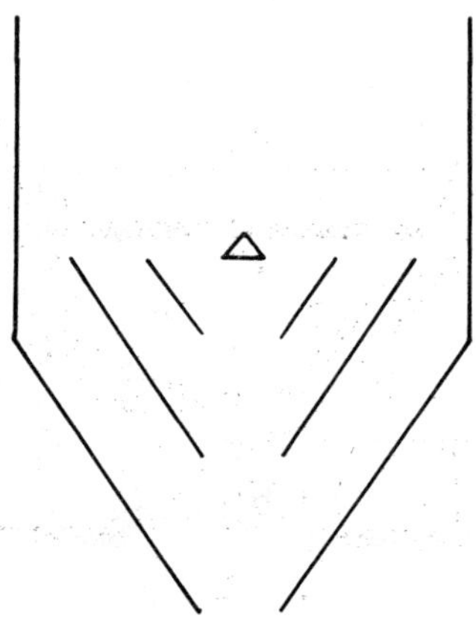

Fig. 65

than the angle of repose of the material, and the number of baffles required will vary with the size of the bin and particle size. The distance between the baffles should be at least three times the diameter of the maximum particle size. This method has been confined to ore that is dry or at least not too wet to be sticky.

Tests with other kinds of baffles to prevent size segregation have also been described by Peacock (223), see Chapter 8.1.

When using a horizontal baffle plate, a cone of stationary material builds up on top. To prevent this, the insert should be constructed as a cone. However, the placement and size of such inserts are critical, for it is important to locate the insert in such a manner that the material cannot form a stable arch between the insert and the hopper, and that - on the other hand - an optimum of active zone may be achieved. Russian authors (91) have found in test models that the flow diminishes or even stops if the distance between the outlet and the bottom of the conical insert is too small. But if the distance is too great, segregation occurs below the insert.

To predict the criteria for the best placement and size of inserts Johanson has developed a theory which was verified by experimental work (138), (141). The test material was a fine grained iron ore concentrate and different cohesions were achieved by adding oil. Very informative relationships between the characteristics of inserts, hoppers and bulk materials are summarized in Fig. 66.

Assuming the slope of the conical insert (Θ_2) to be the same as for the hopper (Θ_1), the criterion for placement and size of the insert is the ratio W/R (W = horizontal distance between bottom edge of the cone hopper wall, R = radius of insert basis). This ratio changes in the diagram with the effective angle of internal friction of the bulk material (δ), with the angle

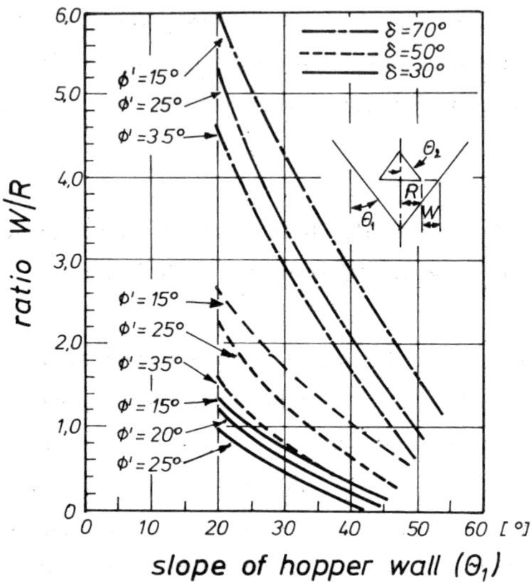

Fig. 66

of friction between wall surface and bulk material (ϕ'), and with the slope angle of the hopper (Θ_1). For example: for a hopper with $\Theta_1 = 30^{\circ}$, angles of friction $\delta = 50^{\circ}$ and $\phi' = 15^{\circ}$, the ratio W/R equals 1.8 by which the size of the cone and its placement is defined. For a material with less flowability, for example with $\delta = 70^{\circ}$, this ratio increases.

When flow around the inserts is fully developed, the pressure on them is less than the peak pressure occuring during initial flow. Therefore, safe structural design is achieved by considering this maximum force. When the insert is used to create mass flow or to prevent segregation these forces can become very large. However, on smaller inserts, which are placed low in the hopper, the forces are considerably smaller.

Johanson and Kleysteuber (140) have theoretically derived and experimentally verified equations which make it possible to predict the loading on the insert. This loading depends on the properties of the bulk material, the bin dimensions and the size and position of the insert. To analyse the properties of the bulk material the above mentioned bin-design theory was used (132), (133), (137). Since the flow channel around an insert is essentially a vertical cylinder, the vertical pressure P in such flowing material is approximated by

$$P = \frac{\gamma \cdot D\,(1 + \sin\delta)}{4 \sin\delta \cdot \sin 2\nu'}$$

γ = bulk density of the solid

D = diameter of flow channel

δ = effective angle of internal friction

$2\nu' = [\sin^{-1} \cdot (\sin\phi' / \sin\delta) - \phi']$

ϕ' = wall angle of friction

The total vertical force exerted on the conical insert is given by the pressure P times the effective area of the insert. This area is simply $\pi \cdot r^2$ (r = radius of the insert) for a free flowing material. For bulk solids with a tendency to arch the flow channel must be wide enough to break the arch. The effective radius of the insert including the arching effect is the distance R (see Fig. 67). B is the width of the largest stable arch which depends on the yield strength f_c of the bulk solid which must be obtained from the strength versus pressure data, see Chapter 2.1.2.

$$B = f_c / \gamma$$

or $$R = r + f_c / 2\gamma$$

Therefore, the total vertical force F on the insert can be calculated as

$$F = P \cdot \pi\,(r + f_c / 2\gamma)^2$$

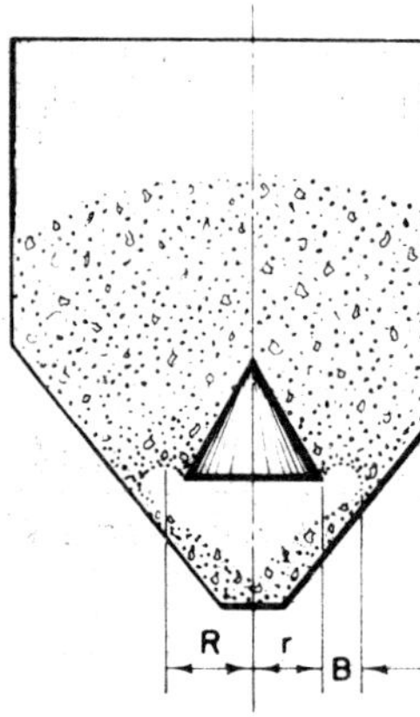

Fig. 67

The authors also give examples for the calculation of forces for bins to store dolomite limestone and sand. Fig. 68 shows possible positions of the structural members for the conical inserts. They should be supported in such a manner that they do not interfere with the flow of material.

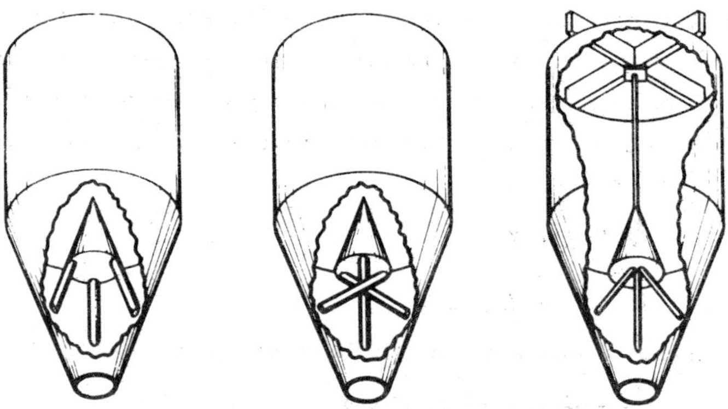

Fig. 68

The advantage of using conical inserts has also been proven by other authors, mostly by fundamental tests (44), (103), (156), (178). By installing an insert, developed by the NCB in Bretby, funneling was eliminated and the volume of mobile material could be increased by as much as ten times (Fig. 69) (7).

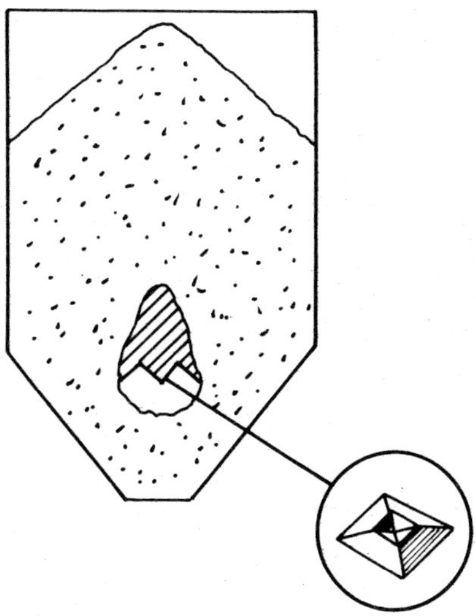

Fig. 69

Photographs of the effect of flow-corrective cones (175) and of wedges which are placed as symmetric and asymmetric obstructions (2 14) give more details of the internal flow of moving beds of granular material. Investigations on the effect of inserts on the flow of crushed limestone in a shaft kiln, which may be considered to act as a bin, were presented recently and will be published soon (99). Several inserts were fitted at different heights to find the best position. The result of each experiment is described by a factor that gives a measure of the equal distribution of velocity referred to its plane in the cross section of the shaft kiln.

Another typical example (202) is mentioned and illustrated in Chapter 8.1. In this case, the combination of inserts and a saddle at the bin bottom made a high flow activity possible which was verified not only visual but also by many series of screen analyses.

Small cones and conical rings in the hopper have also been reported for creating a favourable flow pattern for powdered coal (105). One of the methods to improve flow was developed by the Bitumious Coal Research Inc. It consists of a double-cone which is fitted in a tube below the bunker outlet. This arrangement was found to be superior to the single cone placed within the bunker, especially when handling coal at high moisture content and where the bunker was not operated for several days (75). The whole volume of the tube and the area above it is in motion resulting in a higher discharge rate than that of a similar bunker with a normal bottom outlet (181). Quoting Richards (249): "It was found that compacting pressures in the double cone attachement were low, and thus a low cohesive strength was maintained in the material. Flow through a slot may be more reliable than a flow through a circular aperture, and if one regards the double cone as providing a slot aperture curved round the cone it is less surprising that this device is reported to promote reliable flow".

The latest development of this kind became known as the BCR Easy-Flow-Hopper. Several installations of this kind have been working successfully (60). Fig. 70 shows the shape of this device for coal of minus 6 mm grain size (at left) with up to 12 % and (at right) with up to 15 % moisture content (111). The difficulties which arise from passive zones occur not only in bins and bunkers, but also in rail cars with side discharge for the transport of bulk material. To reduce this passive zone and thus to prevent partly solidifying of material (shaded area Fig. 71 left), Kvapil (164) proposes a saddle in the longitudinal axis of the waggon which corresponds generally in shape to the ellipsoid of motion which was defined in Chapter 1.1.

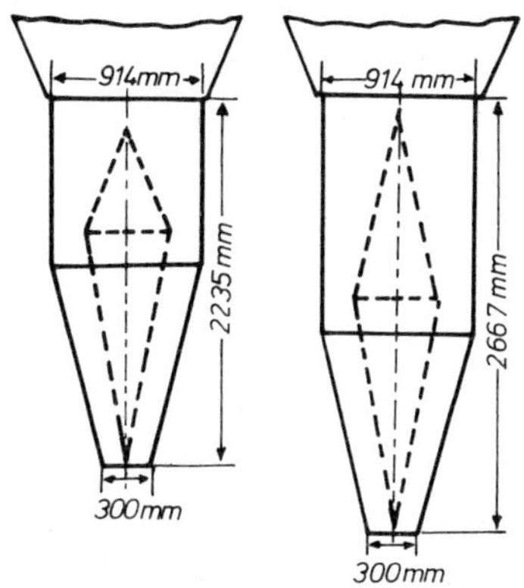

Fig. 70

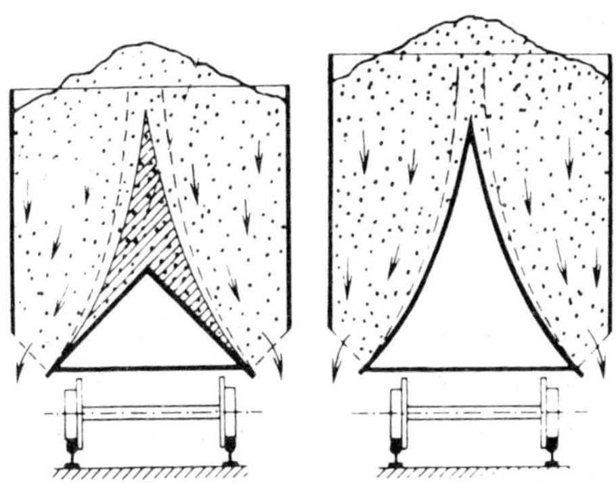

Fig. 71

In some constructions wedge-shaped saddle keys are fitted on the inside wall of bins (F 9), (F 52). Specific positions, sizes and numbers are proposed by the constructors for each storage problem. They are said to be advantageous for bulk materials which tend to arch. As they are arranged in staggered fashions they allow the material to flow with different velocities along the walls. This may also be achieved with one saddle key. They also reduce the pressure of the bulk material acting on the outlet and its discharge devices.

5. Activation of Bins and Bunkers

As mentioned before, the principal design concept for a storage bin is to develop a suitable flow pattern without flow difficulties. These flow difficulties are piping and arching, which have been defined earlier. The active or live storage in the bin should be as large as possible; hence, the volume of moving materials (active zones) should be large. Kvapil (168) has defined the activity of a bin as follows:

$$a = \frac{A \cdot 100}{V} \ (\%)$$

A = volume of the active (moving) zone

V = total volume of the bin

When a = 100 %, the active zone extends over the whole bin, i. e. all material is in motion. Korn (156) defines this activity in similar terms.

In those cases were it is impossible to build bins, bunkers and hoppers in accordance with Jenike´s gravity-flow methods, it may be necessary to use special liners, inserts or special mechanical devices to improve the activation in the bin. This may be necessary for older bins, where the space limitations prevent ideal constructions, or for materials that do not flow uniformly or hardly at all. But it should be considered that improvements to obtain mass flow also may cause a significant increase of wall loads (see Chapters 3. 2. and 3. 3.).

5.1. Liners

One step towards attaining a reasonable good flow lies in the proper selection of the interior wall surfaces in the bin. For this purpose liners or coatings can be used if the original wall surface of the building material does not effect optimum condition of friction with the stored material (42), (43).

In case of using a liner to improve the flow in hoppers, the resistance against abrasiveness must be seriously considered. During bunker operation the liner surface may be altered by the stream of material, i. e. a smooth surface may be roughened, or a rough one may be polished.

The hopper angle should be chosen for the optimum flow as determined by the angle of wall friction, see Chapter 2. 1. 3.

To find the best combination of these factors, Jenike's methods should be considered as long as the space conditions allow it. A good example for the influence of wall liner material on bulk solids flow is given by Kott and Kramer (159) who built a dextrose storage bin according to the Jenike method.

This bin consists of cylindrical, hopper, and transition sections (Fig. 72, left). In order to get mass flow and prevent arching, these sections were coated with different epoxies to get different wall frictions. The roughest wall was in the vertical section, the smoothest in the transition part, a medium roughness in the hopper section. Because of this arrangement sufficient flow was obtained as the flow factor for all bin cross sections was the same. Fig. 72 also shows a pressure profile (right) indicating that three different coatings do not produce as high a consolidating pressure as only one type of coating.

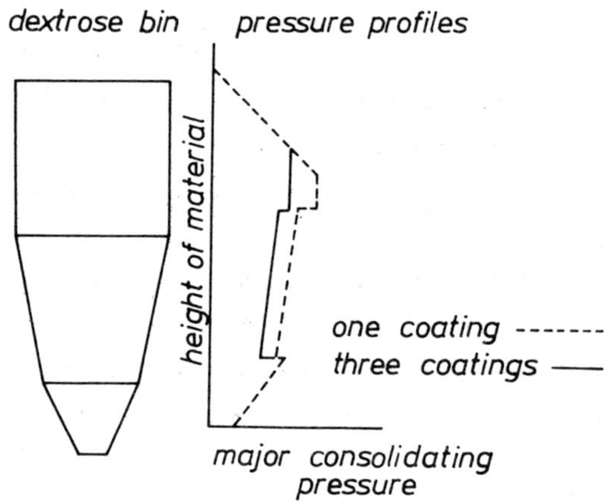

Fig. 72

The influence of hopper walls with different surface roughness on flow rates was tested by Dubach (83), (84). He attached sandpaper on two adjacent sides of square, rectangular and conical test hoppers, while the other two sides remained smooth (steel plate). Comparing the discharge rate with a hopper made of walls from the same material, it was found that the one with two different wall frictions empties faster. For a square or pyramidal hopper the difference was found to be the greatest. The maximum deviation for a free flowing rice was 17,8 %. This surprising result can be explained by the fact, that, because of the different liners and therefore of the different friction angles between the bulk material and the wall surface, the flow velocities along the walls vary, reducing the frequency of forming stable or instable arches.

When storing fine grained bulk materials with a considerable moisture content, such as steam coal with more than 8 per cent, steel plate hoppers or steel plate liners are preferred to other types (251), (329).

Plastic, rubber or hard tile liners as they are mentioned in the chapter on abrasion are also of great advantage to improving flow. In order to avoid repetitions, the reader is referred to that chapter for further information.

Many bulk materials may be considered inelastic in the area where arch stresses occur. If it is necessary to construct a bin or a hopper for these materials, the breaking of an arch may be achieved by using an elastic aperture wall (224). In this case the material will not be compressed, but instead the elastic aperture wall will respond to the increased load by expanding outwards and thus the arch will break. Peschl (224) shows how to calculate the arch stresses and thus the required characteristics of the spring. Such a hopper wall is represented in Fig. 73, where h = height of elastic section, ζ = angle of direction of expansion of aperture wall and ξ = angle of direction when deforming the spring.

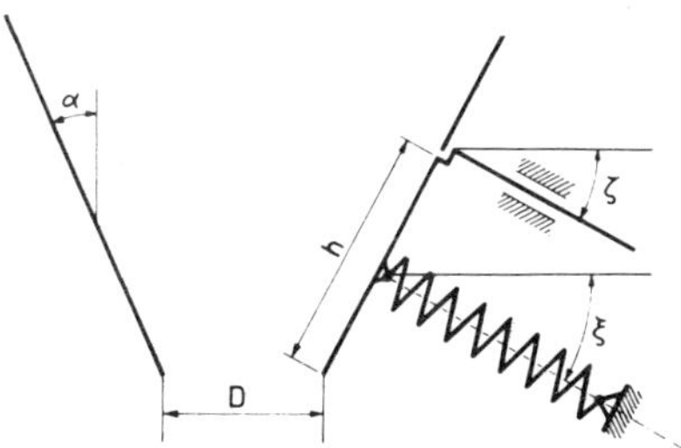

Fig. 73

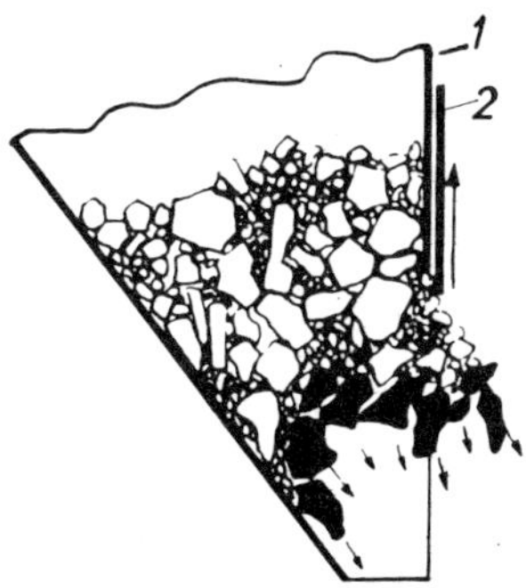

Fig. 74

A simple method to break arches in the lower part of a hopper is shown in Fig. 74 (167). The vertical wall is provided with a movable gate plate (2). In case of arching this plate is pulled upwards until the arch breaks and the stored material is free flowing again. Afterwards the area of outlet is reduced again to its normal size.

5. 2. Flow Inducers

Not all bins and hoppers can be made self-emptying by gravity flow alone and still be reasonably proportioned. Therefore, this chapter will deal with bin-emptying problems where flow by gravity alone cannot be achieved or may be inadvisable. Practical methods to get stored materials flowing or to get a high level of flow activity require the installation of "flow inducers" to the bin and/or hopper. This term, "flow inducer", covers any means or devices that will cause a relative motion between the particles and the wall so that they would slide downwards.

Many types of these flow inducers are being offered and in use. In general all of them are based on one of the four following basic concepts (275), namely

a) pneumatics,
b) vibrators,
c) mechanical dislodgers.

5. 2. 1. Pneumatics

Air may be introduced as a jet in which the kinetic energy dislodges the particles in sudden bursts, or as a gentle uniform flow causing aeration of the material, or as a medium to inflate bunker cushions. Examples will be given for the first and third method in the next paragraphs, while the second method will be described in a separate chapter later. This method is to be outlined together with the principles of aeration (Chapter 6. 2.) which can be considered for discharge.

For loosening the bulk material in hoppers and bins by jets of compressed air with an average pressure of 6 atmospheres (88 psi), air lances are installed through the walls in those regions where arching or funneling may occur frequently. In case of a conical hopper jets are most effective in the area of the transition from the bin to the hopper (219), (292). Fig. 75 illustrates an arrangement for a hopper with one vertical wall. The end of the tube inside the hopper should be bent, so that the compressed air is directed to the discharge opening and can blow the bulk material downwards, restoring gravity flow again. The action of the air lances is regulated by valves, which also may be controlled remotely. A similar method is the so-called puls-jet system (78) and the injection of air through aeration rings suspended in the hopper (52).

An automatic control scheme for a pneumatic jet system is illustrated in Fig. 76 (330). A steel wire 1 is attached to one side of the discharge opening and hold in position on the other side by a counterweight 2. As long as material flows out, the wire is deflected because of the material weight. When flow stops, the wire straightens and lowers the counterweight, tripping a contact which causes compressed air to be blown into the hopper. The air will be cut off as soon as flowing material deflects the wire again.

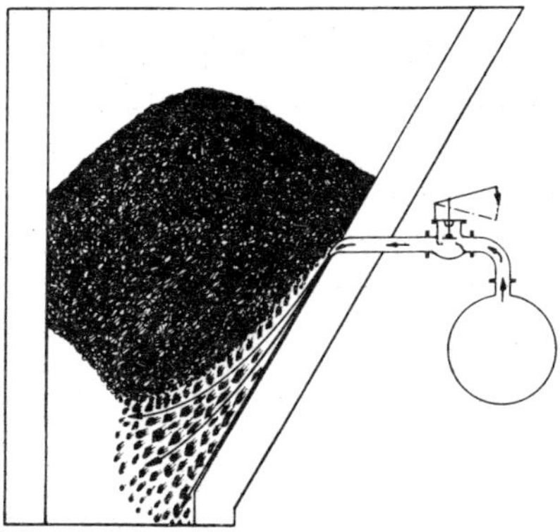

Fig. 75

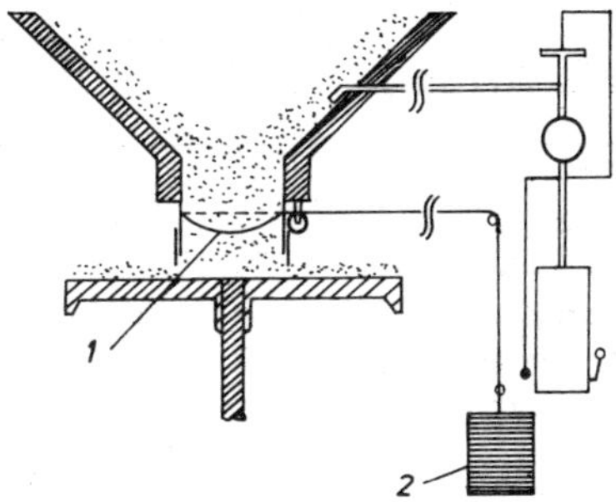

Fig. 76

A special method of pneumatically dislodging a compacted bulk material in bins and hoppers is the Airdox bulk-breaker system (F 35). The components of this system have been applied as an approved, non-explosive method of breaking coal at the working face in underground coal mines. Airdox tubes are permanently mounted on the wall of the bin and extended into the impacted area. They can be discharged at any time to break the material and start it flowing by a pushing and heaving action. The compressor is capable to develop an air pressure up to 840 atmospheres (12,000 psi) and is adequate to serve several bins which are fitted with Airdox tubes. The breaking pressure for most bin applications is in the range of 240 to 420 atmospheres. Absolute control of breaking force is assured by controls built into the system. A schematic drawing of a typical Airdox bulk breaker system for a storage bin is shown in Fig. 77.

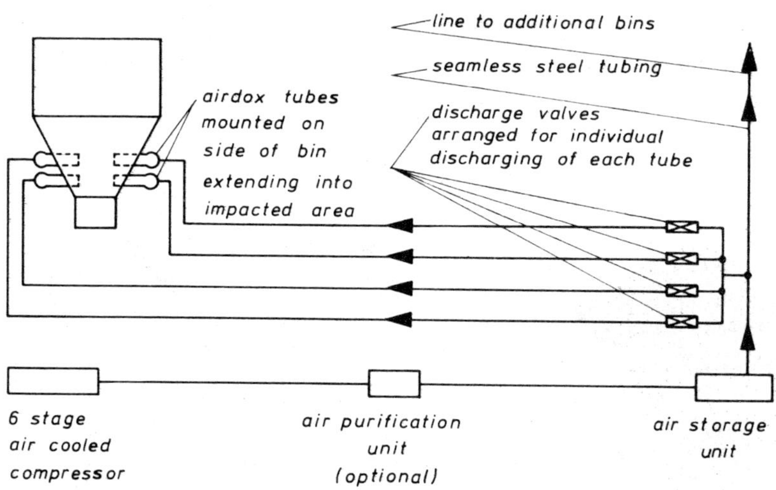

Fig. 77

Another system, based on the expansion of gas at high pressures (700 to 1,300 atmospheres), is the Cardox bulk breaker (F 35). In this case the breaking of compacted material is achieved by using the expansive force of liquid carbon dioxide when it is changed to a gaseous state. This system may be applied for temporary, intermittent discharge. Because of the different working procedure the Cardox bulk breaker requires the tubes to be removed for recharging at a special station.

Compressed air is also used to inflate cushions for safe and rapid discharge of bins, bunkers and silos (181) (F 15), (F 46), (F 47). The cushions from one of the manufacturers consist of approximately 12 mm thick, highly elastic soft rubber membranes, having their edges fixed to an iron base plate. The cushions are inflated by compressed air with a pressure of 3 atmospheres (44 psi), forming a half-balloon and pushing the stored materials, thereby destroying arches and funnels. After flow has been reestablished the cushions will return to their initial deflated position. Inflation time will take 6 to 8 seconds. Although an existing air supply system will suffice, a two-stroke compressor with a capacity of 110 to 1,400 litres per minute and with a working pressure of 6 atmospheres should be recommended for new installations. Fig. 78 gives examples for the arrangement and shape of cushions.

Left: Scheme of a bunker only equipped with rectangular cushions (800 mm x 1,000 mm),

Center: Scheme of a bunker equipped with corner cushions,

Right: Arrangement of a trapezoidal shaped cushion in the hopper of a cylindrical bunker.

With an inflated surface of about 8000 cm^2 the pushing power of a rectangular cushion amounts to approximately 24 tons, which exceeds practical requirements.

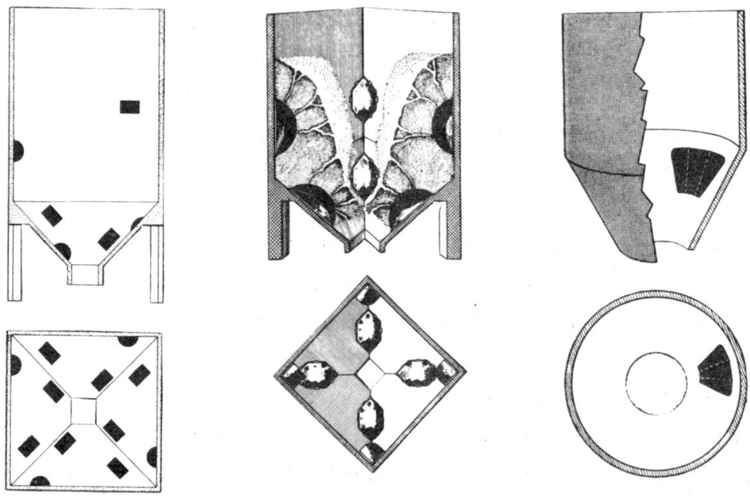

Fig. 78

Automatic control of the whole system of cushions may be achieved by means of sensing contacts near the outlets of the hopper. If there is no flow, the cushions will be inflated.

Most effective position of these cushions is as near to the region of arching as possible (201).

5.2.2. Vibrators

Vibrators as flow inducers are available in many types and sizes, even in dust-tight or explosion-proof construction, and they may be activated by electrical or pneumatical energy. The frequency of their circular or straight line vibrating motions range from 1,000 to 80,000 vibrations per minute and their amplitudes from nearly zero to approx. 6 cm or more (275). They may activate the bulk material by the following methods of installation:

a) units on the outside wall of the bin or hopper,
b) plates on the inside wall of the bin or hopper,
c) suspended units in the bin or hopper,
d) a vibrating hopper suspended from the bin.

Fundamental theoretical considerations regarding the vibrating system (hopper and vibrator) have been published by Taubmann (291). He developed an equation for the elasticity constant which depends on the hopper shape, the thickness of the plate, and the kind of stiffeners.

5.2.2.1. External Vibrators

The basic requirement for the application of an external vibrator is a bin or hopper wall capable of vibrating. Each point of the wall will vibrate with a different amplitude and the wall represents an oscillation system of given mass and natural bending frequency. The knowledge of these values is of advantage in the determination of the optimum vibrator. While the mass usually can be calculated or simply estimated, the irregular shape, varying degree of rigidity of the wall and the usual presence of stiffeners do not normally permit a mathematical calculation of the natural frequencies. Thus these conditions are usually tested by trial and error (F 1).

Vibrators should operate only while the hopper is open to flow, otherwise they may pack the hopper contents (F 22), (F 58).

The sketches in Fig. 79 show suggested application points of vibrators on various types of bins and hoppers (F 23). Although the examples are given for only one vibrator, several vibrators can be used at opposite faces and at different elevations.

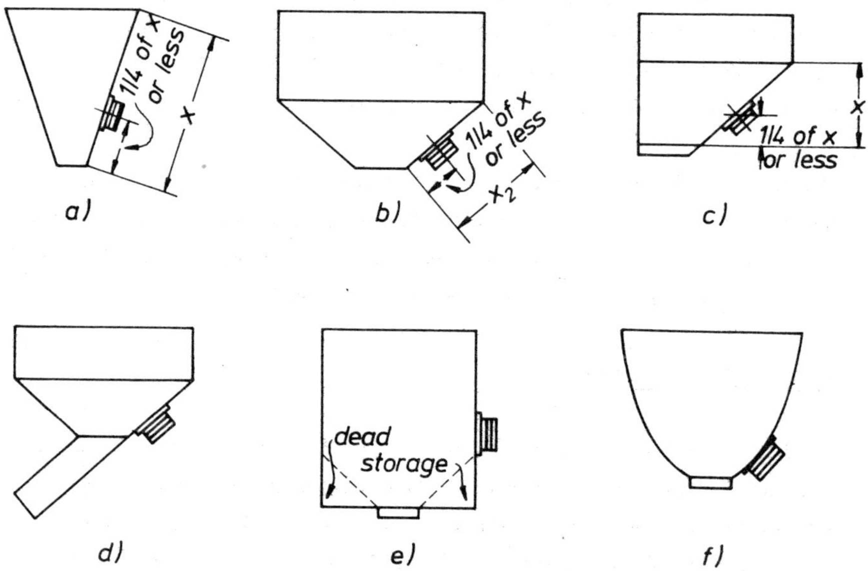

Fig. 79

a) Conical hoppers: The vibrator should be bolted directly to the hopper face one third to one fourth of the distance from the discharge to the top. Rectangular hoppers: The same rules should be applied. If there is a stiffener the vibrator should be mounted in the middle of the panel adjacent to the stiffener.

b) Rectangular bins with hopper bottoms: The vibrator is mounted in the same relative position as for the rectangular hopper. Generally a slightly larger vibrator is required due to the superimposed load.

c) Hoppers with vertical side: The vibrator should be placed on the sloping face directly opposite the vertical side. The location in regard to height is the same as before.

d) Hoppers with sloping discharge spouts: The position of the vibrator should be on the bottom line as shown and as close to the discharge spout as possible. If necessary additional vibrators can be used on the discharge chute.

e) Rectangular or cylindrical bins with flat bottom and center discharge: The vibrator should be mounted directly to the side of the bin just above the point where the angle of repose intersects the sides.

f) Parabolic hoppers: The vibrator should be bolted directly to the plate within approx. 20 to 25 cm of the outlet frame.

Electromecanical rotary vibrators should use a channel mounting as shown in Fig. 80 (F 23). This channel section distributes the vibration throughout

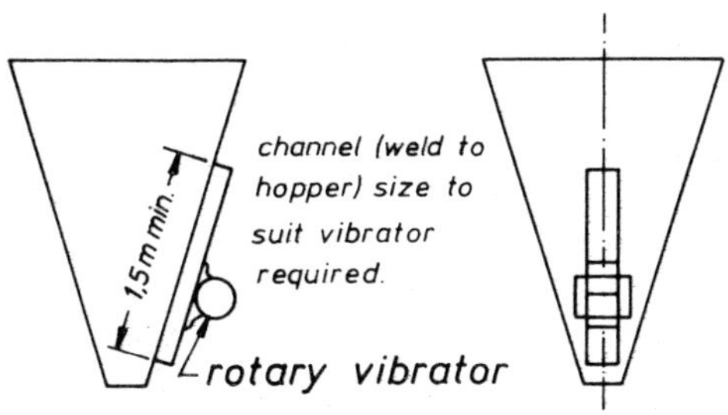

Fig. 80

the bin or hopper, prevents damage and reinforces bins and hoppers of light construction. The length of the channel should have a minimum of 1.5 m.

A special application of portable vibrators is represented by car shakers. They can be hung from hooks or brackets on railroad hopper cars containing coal, sand, gravel, sugar, flour and other bulk materials. These light weight vibrators can be handled easily. Users say unloading time is cut by 30 per cent and the hazards connected with climbing around the railroad cars and poking into its contents are eliminated (F 14).

A vibrator can also be used to control the type of flow in a hopper (Fig. 81) (181). An unsymmetrical filling (left) may result consequently in unsymmetrical loading and may cause an overflowing before the hopper is filled completely. Moving the vibrator in the position as shown at right, an acceleration of particle flow along that particular wall was achieved so that the top of the bulk material is leveled.

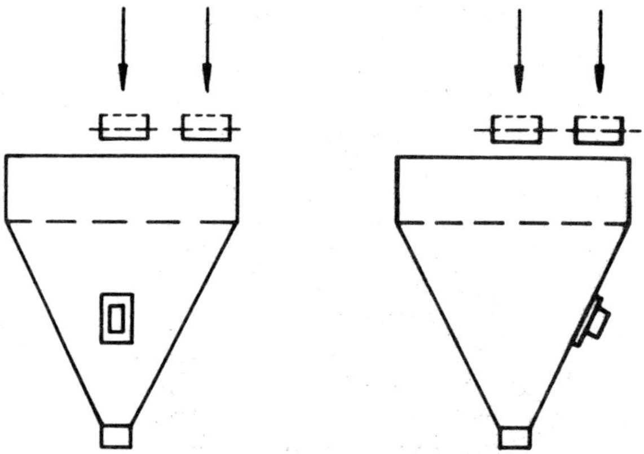

Fig. 81

Andrews (19) discusses the effectiveness of a vibrator as influenced by its position, the design and position of stiffeners. To obtain the desired mode of vibration for best flow, he proposes to place horizontal stiffeners at a specified distance from the outlet, to achieve a degree of stiffness in connection with vibration frequency and intensity, S h i n o h a r a (274) has found among other relations that the vertical sinusoidal vibration of lower frequency or greater intensity gives a larger impact velocity of powder bed against the hopper wall. This is effective in the discharge of cohesive materials from a blocked hopper (145), (208), (307).

There are two different types of external vibrators:

electromagnetic vibrators and
rotary eccentric vibrators.

The second type may be driven electrically or by means of compressed air.

Generally, an electromagnetic vibrator has a longer working life and is simpler than the rotary vibrator. It produces a vibration vertical to the bunker wall and has no rotating parts which are subjected to wear and tear. The rotary vibrator, on the other hand, imparts a radial impulse, so that the bin or hopper are also subjected to stresses parallel to the wall. In spite of these disadvantages rotary vibrators are suitable for moving particularly heavy materials out of bins and hoppers having reasonably thick walls. Most of them are designed for intermittent operation. Therefore it cannot always be decided in advance which type of vibrator may be most advantageous. As a guide Table 6 lists some conditions affecting the choice of type of vibrator (F 23).

If a larger bin has several electromagnetic vibrators mounted on different walls, it is desirable that they should operate with a phase displacement. Only when several vibrators are arranged next to each other on one flat wall, the connections can be made to the same phase.

Table 6

application	electro-magnetic vibrator	rotary-eccentric vibrator
linear vibration, perpendicular to bin or hopper surface	x	
non linear or slightly rocking action		x
greater amplitude, or stroke, i. e. on damp materials		x
higher frequencies, small hoppers	x	
higher frequencies, large hoppers	x	
low noise level		x
more than one vibrator per bin	x	
intermittent operation	x	

The size of a vibrator required to promote the flow depends on many elements, such as size, shape, construction (inclination of the hopper, thickness of hopper plate, number and stiffness of supports), material of the hopper, and conditions of material. The grain size has a great deal to do with the flowing of material. As mentioned before a coarse material generally is easier to move than a fine powder and thus requires considerably less vibration.

Though it is not possible to set definite rules for the size and application of vibrators, some relations exist between vibration frequency, maximum thickness of the hopper wall, and approximate capacity of the bin (F 23) (Table 7).

Table 7

thickness of metal wall (mm)	approx. capacity (t)	electromagnetic vibrator		rotary vibrator (electromechanic)	
		vibrations per minute	unbalanced weight (kg)	vibrations per minute	unbalanced weight (kg)
8.0	5	3600	16	3600	17
9.5	20	3600	30	1800	25
12.7	50	3600	95	1200	71
50.8	150	3600	426	900	260

Relations between some other factors to determine the most effective electromagnetic vibrator are also set up by Nakamura (208), (209) as shown in Table 8.

Table 8

bin no.	1	2	3	4	5	6	7	8	9
approx. capacity of the bin (t)	0.1	0.35	1	3	20	20	50	150	150
thickness of hopper wall (mm)	1-1.6	1.6-3.2	3.4-4.5	4.5-6	8-10	8-10	10-13	13-25	25-40
input of the vibrator (watts)	75	75	150	250	400	750	1500	2200	3000
max. thrust of the vibrator (kg)	40	70	150	250	500	800	1600	2000	2500
unbalanced weight of the vibrator (kg)	7	7	13	18	27	43	93	110	115

Based on these recommendations, the same author also gives, in Table 9, examples for the successfull storage and flow of several materials when using a vibrator as activator for the conditions and combinations given in Table 8.

Table 9

material	moisture content (%)	grain size (mm)
iron sulfide	10 - 20	0.1
calcium carbonate	-	0.05
magnesia	7	0.1
fluorspar	10	0.8 - 0.15
lake pigments	-	0.7
limestone	5 - 15	0.5
raw material for firebrick	3	0.15
soil	8 - 20	1.1
power plant coal	7	0.07
blast furnace dust	10	0.15
sawdust	50	-

Electromagnetic external vibrators with 6000 vibrations per minute are mainly used for smaller and thicker walls with stiffeners, so that they have higher characteristic frequencies. As curved walls are considerably more rigid than flat ones, vibrators of 6000 vibrations per minute are used exclusively (F 1).

By using rotary-eccentric vibrators with a higher frequency, such as 10,000 to 20,000 vibrations per minute, it is possible to generate improved flow at low amplitude and to remain within the elastic limit of the hopper walls, which will avoid a fatigue of material. Thus small sized vibrators with small amplitudes but high forces will get material out of hoppers, which could not be emptied by anyother method. For instance, a turbine vibrator of only 20 kg of unbalanced weight will generate a force of 6 tons at 14, 500 vibrations per minute. These are air operated vibrators which run without bearings due to their construction. Besides turbine vibrators there are also pneumatic roller (Fig. 82, left) or ball vibrators (right),

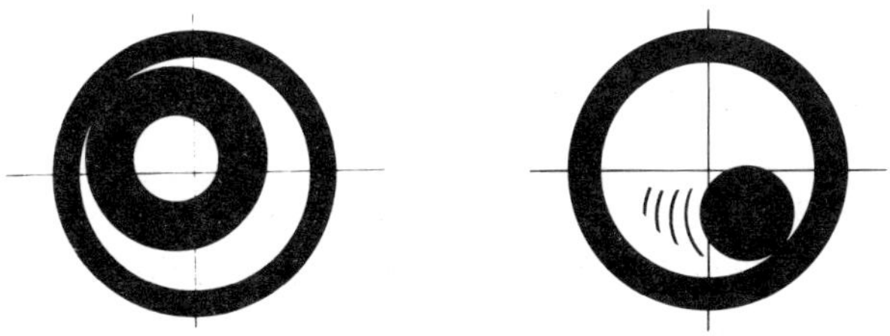

Fig. 82

also without bearings. All these models have adjustable speed and force by simply changing the air pressure. The maximum force for ball vibrators is approx. 400 kg, for roller vibrators 1, 100 kg, for turbine vibrators 6, 000 kg (F 44).

Another method to generate vibration to a bin or hopper wall is by air-powered piston type vibrators. Depending upon the type they may be applied for bins and hoppers up to 100 tons capacity and more and the vibrations are variable between 1, 000 to 11, 000 per minute. Fig. 83 shows at left side a sketch of a vibrator with air-cushioned piston, which is ideally suited for low-noise level installations where the material flow characteristics are not too stubborn. The vibrator on the right hand has a metallic impact piston which is preferred for bulk materials with poor flow properties (F 14).

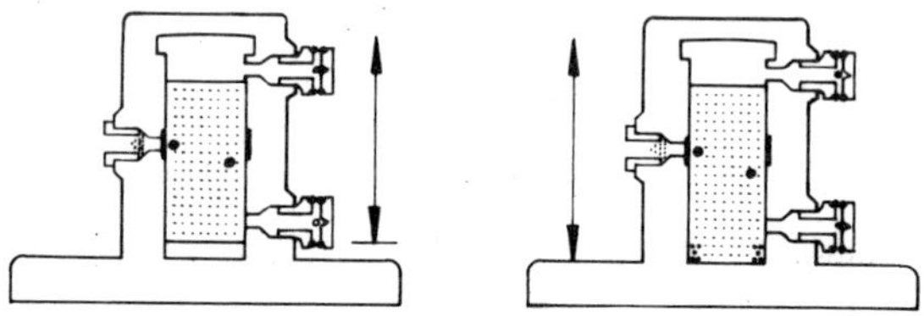

Fig. 83

5.2.2.2. Vibrating Plates

In the case of rigid bunker walls, such as in concrete bunkers, additional vibrating plates can be mounted on the inside wall of the bin or hopper.

When using an electromagnetic vibrator the following construction is recommended for the hopper (F 23): A steel plate, the size of which is determined by the different storage conditions, should be placed above the discharge opening along the hopper side and mounted with the vibrator. The plate should be completely isolated from the hopper wall by rubber absorbers. The vibrator should be attached at approximately one quarter or less of plate height from the lower edge of the plate. Connection is by bolts going through the hopper wall.

Another example shows the application of oscillating slide plates with rotary vibrators. The shapes of the plates may be rectangular or trapezoidal, so that their shapes equal as much as possible the area which is to be vibrated. A trapezoidal shape is suitable for a hopper wall of a square or rectangular bin cross-section. Further explanation is given in Fig. 84 (F 19), which shows different methods of installing vibrating plates in concrete hoppers. In type A, there is an opening in the hopper wall large enough to insert the

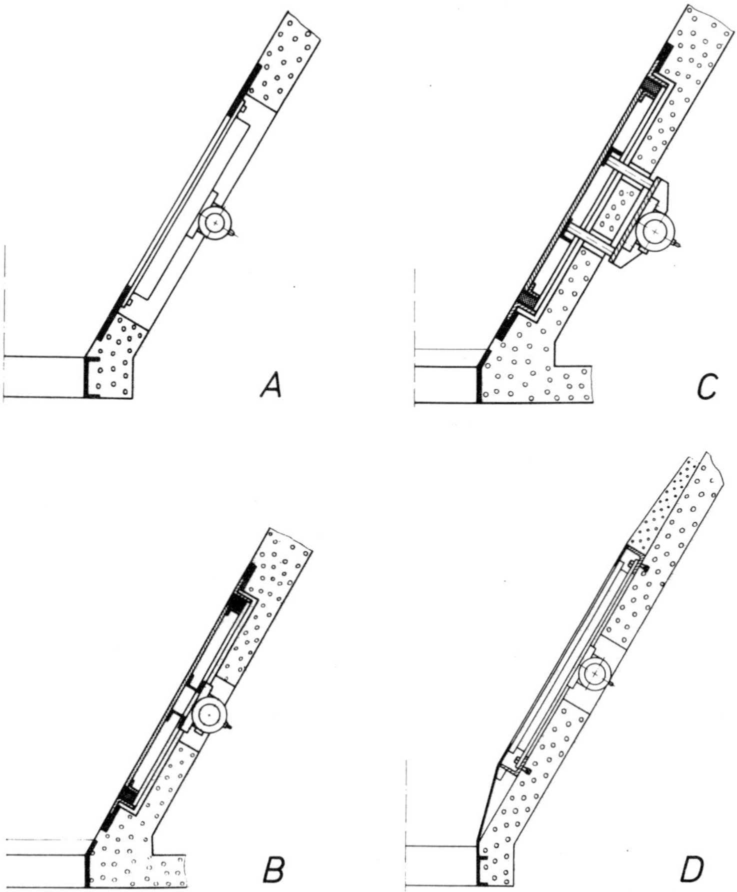

Fig. 84

complete vibration unit. It has been used in a bin for the storage of gypsum in a cement plant. In type B, the plate is mounted in a depression of the wall, while the vibrator itself is placed through a hole in the wall. This arrangement has been used for coal in a power station. Type C shows the connection between the plate and the vibrator by means of bolts, a similar method to the one mentioned in the previous paragraph using electromagnetic vibrators. The easiest way to install a vibrating plate in an already existing hopper is to mount the plate on the wall and minimize the obstruction by providing a smooth transition from the wall to the vibrating plate (type D). For the vibrator a separate hole in the wall is required.

For some powders it has been found helpful to pass curved steel rods through the bin wall penetrating 25 to 30 cm into the material (249) and connect them to vibrators.

5. 2. 2. 3. Suspended Vibrating Elements

Sinden (275) gives a typical example of a vibration which was applied to a vertically suspended steel plate in the center of a bin (Fig. 85). By this

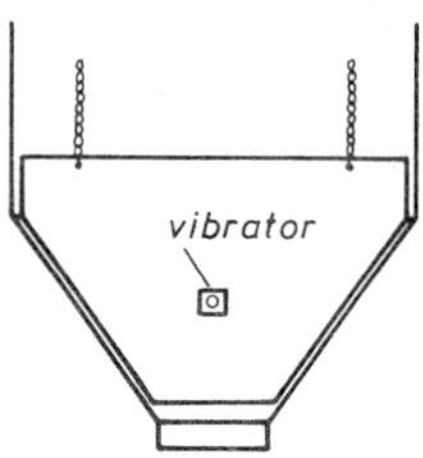

Fig. 85

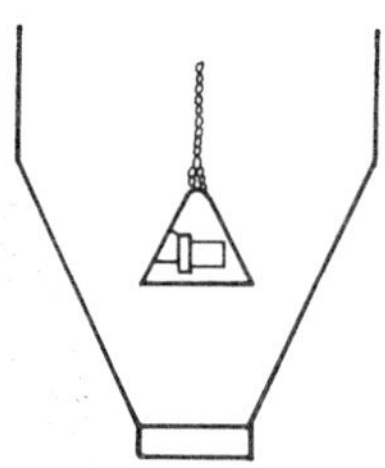

Fig. 86

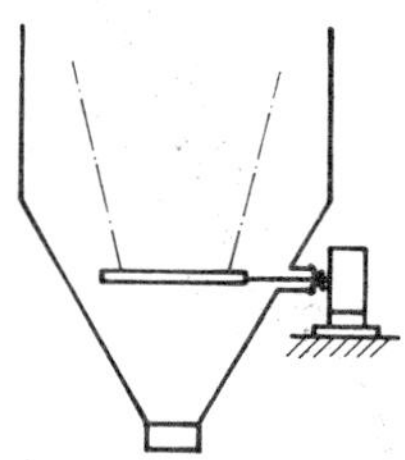

Fig. 87

suspended arrangement it is possible to tolerate a much greater amplitude of vibration than in case of vibrators fixed to the wall. Since only a small portion of the vibration is transmitted to the wall the danger of damaging the bin construction is reduced considerably. Suspended members can be recommended for bulk materials which require a great amplitude of vibration for better flow. The above mentioned example has shown that the effectiveness of the vibrating member was considerably greater than that of vibrating hopper walls.

A similar vibration may also be achieved by a vibrator in a suspended casing as shown in Fig. 86 (260). An additional advantage of this arrangement is given by the slow swinging motion of the suspending cable or chain.

Poulton (233) describes a dish-shaped vibrating unit with a circular convex baffle which is suspended in the bin by rods. By the vibrating internal baffle a shearing action is produced in the material, thus promoting flow through the annulus space.

A vibration within the bulk material may also be induced by suspended gratings which are activated by vibrators from outside the hopper wall (Fig. 87) (F 58). These gratings are made from steel and their sizes depend on the cross-section of the hopper. Very good results have been achieved in stored salt, chalk, titanic oxide, sulfur, powders of lime, quartz, rubber, sawdust, etc.

5.2.2.4. Vibrating Hoppers

The use of vibrating hoppers for bin activation is increasing and their application methods are improving a wide variety of materials. A vibrating hopper must be isolated from the storage bin and the supporting structure in order to avoid damage. This is accomplished by a flexible connection (C) between the vibrating hoppers (D) and the stationary section of storage bins (B) (Fig. 88). Through this vibration the flow pattern in the bin is substantially enlarged (A) and equals in its diameter to the inlet of the vibrated hopper section.

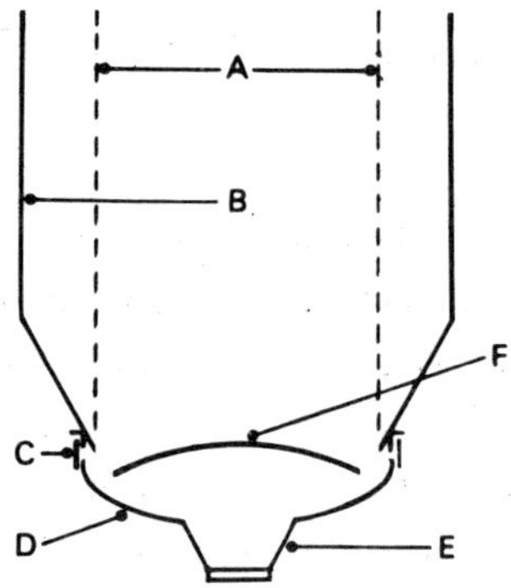

Fig. 88

The bin activator shown in Fig. 88 (F 54) incorporates essentially a small funneled outlet from the activated section (E) and the baffle (F) which supports the head load of the stored material and imparts the vibrations directly into it. Vibrated at a frequency of 1,500 to 1,600 cycles per minute many kinds of dry and semi-dry bulk materials with a poor flowability are handled successfully, avoiding the problems of arching, funnel flow, flushing, spasmodic flow and segregation. The stored material thus moves through the bin on the "first-in/first-out" principle. The activators can be fitted to any size of hopper up to 500 tons capacity of the total storage bin. They can also be installed in existing bins which are to be modified.

According to Myers (205), (206) there are two main types of vibrating hoppers. The first utilizes, like the one described above, one excitor mechanism and is referred to as a gyrated-type hopper (Fig. 89). It has eccentric weights which produce a vibration perpendicular to the flow channel. The other type is characterized by two vibratory motors mounted at an angle of 180 degrees apart in such a manner that a lift-twist action is exerted on the

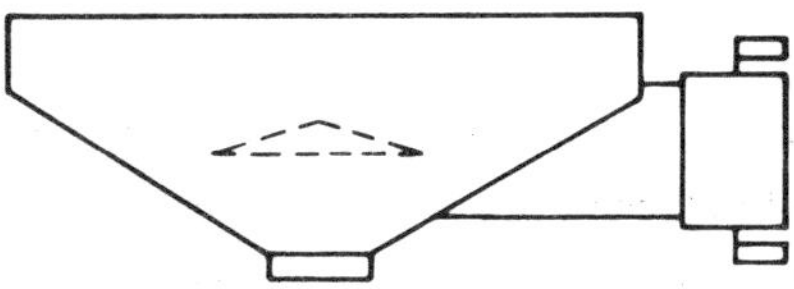

Fig. 89

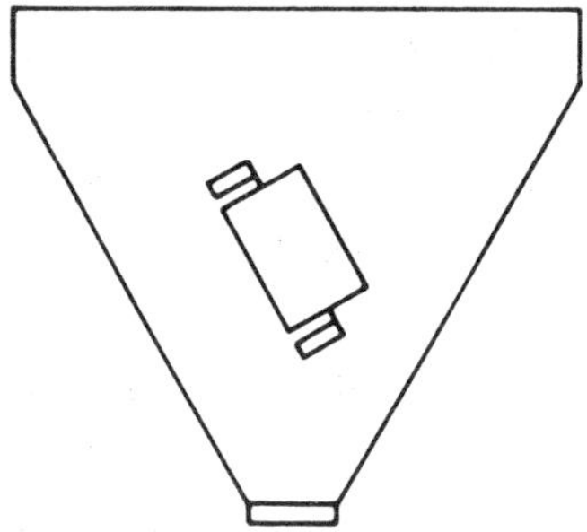

Fig. 90

material. This construction is called a whirlpool-type hopper (Fig. 90) (85) (F 12).

Both types are equipped with internal pressure cones which allow the material to converge at the outlet and, besides this, produce a larger flow stream from the upper bin.

For dry bulk materials of fine or large grain size the gyrated hopper is constructed with a wall slope of generally 30 degrees. With steeper hoppers up to 60 degrees (to the horizontal) this type also discharges successful when storing power plant coal of approximately minus 0, 3 mm size and 15 to 20 % moisture content. When handling sawdust the vibrator hopper should have at least the same diameter as the storage bin, especially if the material is relatively wet.

The whirlpool-type hopper, which is more expensive, has the great advantage that it will handle a wide variety of materials up to extreme moisture contents. It is utilized with a 60 degrees cone. A sawdust, for example, with 70 % moisture can be stored and discharged without difficulties. The lightest material being handled in this hopper type is carbon black, weighing only 43 kg/m^3, and the heaviest is magnetite with approximately 3200 kg/m^3. - Further examples of materials successfully discharged by vibrating hoppers are quoted by Stepanoff (281) in reference to plant location, size of bins and products handled.

Another type of electromagnetic vibratory feeder for powders and granules using a different method for discharging is shown in Fig. 91 in three different positions. The discharge tube of the hopper empties into a housing.

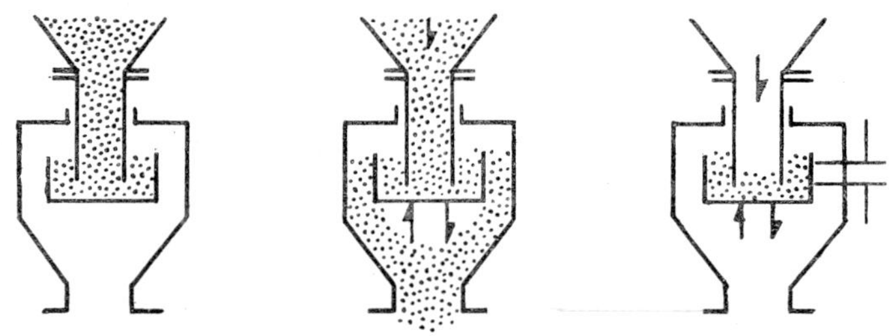

Fig. 91

The important device consists of the circular pan below the bottom edge of the tube, which prevents overflow of the material (left). As the plan is connected to a vibrator, the armature oscillates in a helical movement and the material flows from the feed tube to the periphery of the armature and falls into the outlet chute (center). The flow rate may be controlled by the distance between the feed tube and the armature, the height of the edge ring, and the amplitude of vibration (F 30).

Vibrating hoppers are also installed under large outdoor storage piles (Fig. 92). To increase the vibration effect into the bulk material and thus to enlarge the flow pattern, the hoppers are normally constructed with

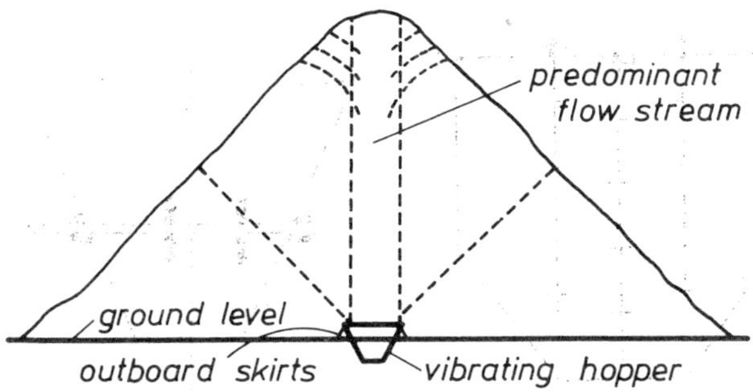

Fig. 92

special outboard skirts. The drawdown angle approaches the angle of repose of the material (205), (206).

5.2.3. Mechanical Dislodgers

One of the simplest methods of mechanical dislodgers is to provide the bin and hopper with poke-holes through which the bulk material may be loosened by means of rods in case of arching or funneling. This method is just as bad as pounding the walls with hammers to promote flow, for their effect is mostly negligible and damage can be caused. Steady control by personal is necessary. Therefore mechanical equipment should be preferred. The literature refers frequently to vertical or horizontal revolving stirrers, rodding devices, rotating paddles etc. in many variations. Two examples are illustrated in Fig. 93 (215). Most of these installations do not work successfully as considerable repairs are necessary expecially when storing hard and large-sized materials. A more successful method for very large lumps of crushed rock or ore is to hang a long chain down the center of the bunker. This chain may be lifted a few feet by a crane to destroy arch formation.

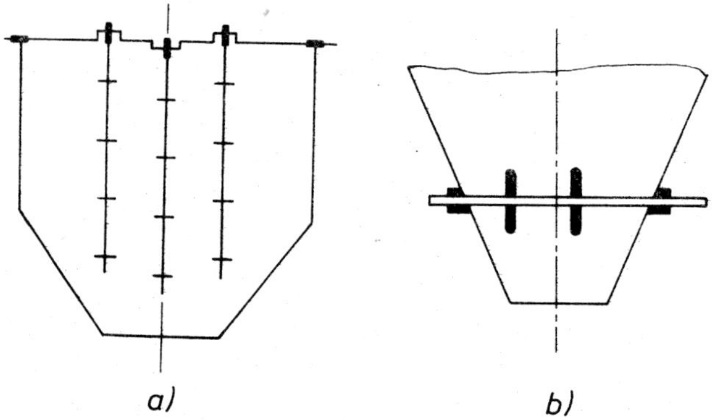

Fig. 93

But there are still considerable problems involved in fastening the chain (249), (263).

For discharging powdered, granular or small lumpy materials with very poor flow properties, circular bin dischargers are used in a growing number (F 49), (F 55).

Sinden (275) has shown the superiority of this device over other methods for very difficult materials. It consists of a circular bin discharger (see Chapter 6.1.4.1.) on which a vane-type arch-breaking arm is fitted by a universal joint. The archbreaker which rotates with the feeder is free to work anywhere within the conical bin. The top of the arm is fitted with a steel spider which may contact a wear bar in the bin (Fig. 94) (F 55).

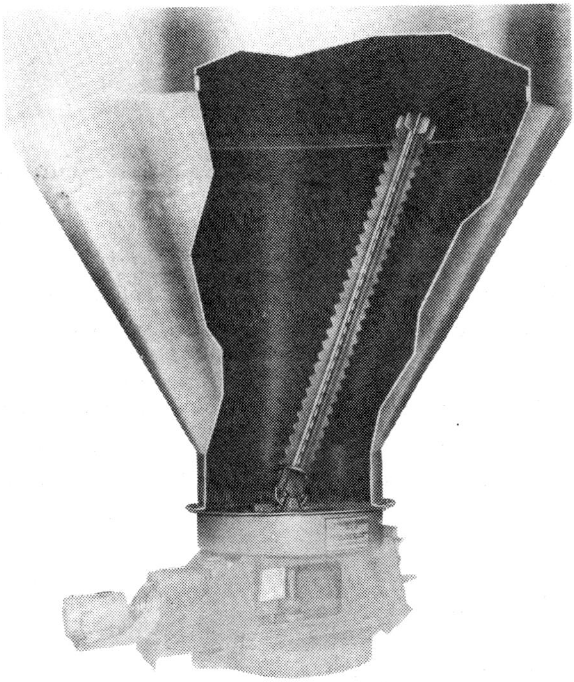

Fig. 94

6. Feeding and Recovery[x)]

6.1. Mechanical Discharge Arrangements

Feeders are used to provide a means of control for the withdrawal of bulk materials from storage units, such as bins, bunkers, silos and hoppers. This control function can only be performed properly as long as the bulk materials flow by gravity to the feeder in a uniform and uninterrupted

----

[x)] Co-author: H. Colijn, Consulting Engineer, Monroeville, Penn., USA fashion. A feeder can do many things, but it should never be considered as a suction pump.

Feeders must be considered as an integral part of the overall bin and feeder system. Improper design of either one of these parts will affect the performance of the whole system. Much effort has been devoted in the past decade on the development of functional design criteria for hoppers and bins and this subject is covered in Chapter 2. The integrated concept of bin and feeder design requires quantitative analysis of the bulk material characteristics before attempting the design and selection of the components.

The design of a feeder system must start first with the proper dimensioning of the hopper outlet to prevent arching, doming or ratholing. The hopper opening size should be large enough to allow passage of the bulk solid at the required maximum discharge rate. A feeder can only throttle the flow.

6.1.1. Feeder Loads

During the past ten years various efforts have been made to determine accurately the load or pressure on feeders mounted directly underneath the hopper opening. For many a designer this pressure still equals the "hydrostatic" head of material above the opening. Consequently, to eliminate this high pressure, he tends to locate the feeder in an offset position from the hopper opening and connects them by way of a spout.

Fig. 95 illustrates three examples of how the bin load may act on the feeder:

a) the full load (not equal to the "hydrostatic" head of material) acts on the feeder,
b) the load is partly reduced by changing the shape of the hopper,
c) the load is completely removed from the feeder and acts only on the hopper wall.

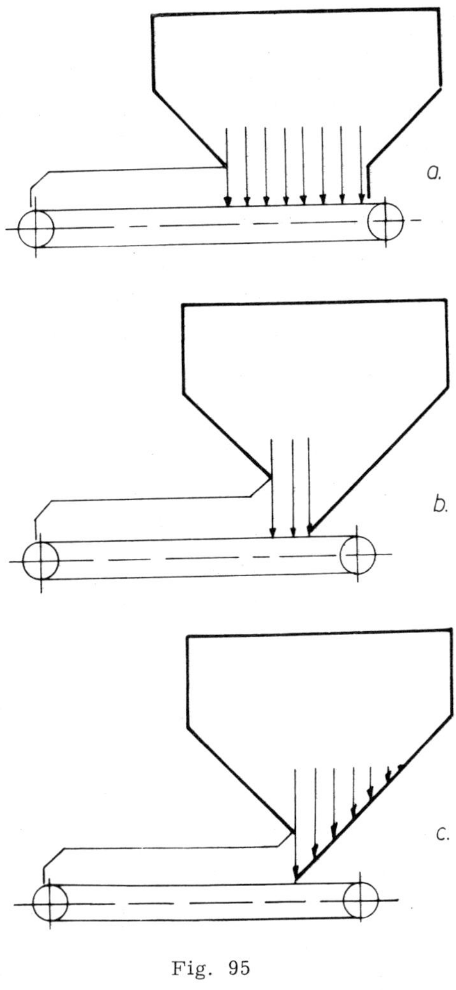

Fig. 95

Although the advantages of b and c appear obvious, one must consider that the effective outlet area is reduced which may influence the flow pattern of the bulk material. Therefore the final choice must be related to the material characteristics.

The offset position of the feeder becomes an important advantage when the bin must be completely emptied before it is filled again. Thus, any possible damage of the feeder by the impact of the bulk material is reduced (92).

Recent studies (70), (113), (133), (137) however, have shown that greatly reduced pressures occur at the bottom of hoppers, because of the converging configuration of the hopper walls. Jenike has developed analytical equations (133) to express the feeder-load in terms of hopper geometry and material properties such as friction angles, shear strength and densities.

From current experimental work on feeder-loads (70), (113) it appears that Jenike's equations tend to compute too low a feeder pressure under steady flow conditions. Jenike's analysis of feeder loads is based on the vertical component of the stress that occurs within the bin. His equation for feeder load during flow is $Q = q \cdot \gamma \cdot B^3$ for circular outlets and $Q = q \cdot \gamma \cdot LB^2$ for rectangular outlets, where γ is bulk density, L and B are hopper outlet dimensions and q is a factor dependent on the hopper geometry and material properties. He provides various graphs (133) to determine the values of q.

Confronted with the higher experimental values of feeder loads, one should consider strongly the possibility that the stress field shifts and rotates above the feeder, causing a different vertical stress condition. The highest value would occur when the major consolidation pressure would act downwards. This major consolidation pressure (σ_1)[x)] can be expressed by

$$\sigma_1 = \frac{\gamma \cdot B \cdot ff}{m + 1}$$

----

x) Notation used is the same as in Jenike's publications

where

γ = bulk density

B = width of hopper opening

ff = flow factor

m = shape factor (m = 1 for circular opening and
m = 0 for slot opening)

The bin wall pressure (σ) at the opening is expressed by $\sigma' = (c)\ \sigma\ B$ where (c) is a factor dependent on hopper geometry and material properties, also available in graph form in Jenike's bulletin (133).

These equations are compared with actual measured feeder loads of two experimental installations. The installation of the University of Hannover (113) was used in 1964 to determine pressures on feeders during filling and discharge of the bin. The second installation was used by Syntron (F 13) during an experimental testing program in 1967.

The comparison of the experimental data and the computed unit feeder pressures are shown in Tables 10 and 11. The last two columns in these tables show the range of feeder loads that the authors encountered in the industry. There are still manufacturers who assume hydrostatic pressures on the feeders. Some manufacturers consider a three-foot column of material on the feeder as the feeder load ($3\ \gamma$). This last method is acceptable for opening widths of 2 to 3 ft, but gives too low a value when 6 ft wide feeders are concerned.

Table 10 indicates that the wall pressure equation $\sigma' = (c)\ \gamma\ B$ computes a unit feeder pressure which corresponds fairly well with the measured pressures. Table 11 shows generally a higher feeder load than the calculated wall pressures at the opening. This may have resulted from additional compaction of the material due to the vibration of the feeder.

Table 10: University of Hannover tests. Belt feeder load in psf under bin-opening during flow

materials	Hannover tests avg. values	Jenike $Q = q \gamma B$	max. consol. pressure $\sigma_1 = \frac{\gamma B ff}{m+1}$	wall pressure $\sigma' = (c) \gamma B$	range of industry calculations	
					hydro-static	3γ
iron ore fines (-4 mesh) $\gamma = 135$ moisture: 2.57 % $\phi' = 30^o$, $\delta = 50^o$	245	95 q = 0.32	350 ff = 1.2	220 (c) = 0.75	3,250	405
coarse iron ore (+1/4 in. to - 3 in.) $\gamma = 155$ moisture: 0.35 % $\phi' = 25^o$ $\delta = 40^o$	285	125 q = 0.38	450 ff = 1.35	320 (c) = 0.95	3,700	465
dolomite stone (- 2 in.) $\gamma = 97$ moisture: 1.68 % $\phi' = 25^o$ $\delta = 45^o$	175	70 q = 0.33	272 ff = 1.3	190 (c) = 0.90	2,300	291

note: material properties were estimated, B = 2.27 ft

Table 11: Syntron vibratory feeder tests. Net feeder load in psf under bin-opening during flow

materials	Syntron tests avg. values	Jenike $Q = q\gamma B$	max. consol. pressure $\sigma_1 = \frac{\gamma B ff}{m+1}$	wall pressure $\sigma' = (c)\gamma B$	range of industry calculations hydro-static	3γ
Lac Jeannine concentrate (specular hematite) $\gamma = 185$ moisture: 1,5 % $\phi' = 25^0$ $\delta = 37^0$	280 (6^0 down slope) 250 (level)	86 q = 0.40	300 ff = 1.4	236 (c) = 1.1	2,600	555
taconite pellets $\gamma = 130$ moisture: 0 % $\phi' = 18^0$ $\delta = 33^0$ (estimated)	270 (6^0 down slope 260 (level)	76 q = 0.50	256 ff = 1.7	212 (c) = 1.4	1,820	390
limestone (- 2 in.) $\gamma = 100$ moisture: 4 % $\phi' = 25^0$ $\delta = 45^0$	170 (6^0 down slope) 155 (level)	38 q = 0.33	150 ff = 1.3	105 (c) = 0.90	1,400	300
Mesabi ore fines (- 1/4 in.) $\gamma = 140$ moisture: 7 % $\phi' = 30^0$ $\delta = 50^0$	160 (6^0 down slope)	52 q = 0.32	194 ff = 1.2	122 (c) = 0.75	1,820	390

note: B = 1.16 ft.

Two positions of the vibratory feeder are listed; 6^{0} down slope and level. It appears that the level position has a slightly lower feeder pressure than the 6^{0} down slope.

The experimental data tend to support the contention that Jenike's equation $Q = q\,\gamma\,B$ for a slot opening results in too low a feeder pressure. Therefore, it is suggested that the wall pressure equation $\sigma' = (c)\,\gamma\,B$ be used for calculating the feeder load of belt-, apron-, and table feeders. For vibratory feeders it appears more appropriate to use the major consolidation pressure

$$\sigma_1 = \frac{\gamma \cdot B \cdot ff}{m + 1}$$

These equations reflect only the feeder loads under steady flow conditions. For design purposes it is necessary that allowance be made for the static load on the feeder during filling of the bin.

During the filling of the experimental bins the pressure on the feeder increased rapidly and then approached some asymptotic value, see Fig. 96.

The filling pressures on the feeder are usually two to four times as large as the steady-state-flow discharge pressures of the feeder depending on point of impact and filling rate. If 8 to 10 ft of height of material is left in the bin when the feeder stops and the bin is refilled, then the filling pressure rises only 10 to 20 % above the discharge pressure. It is therefore recommended to avoid impact of material directly over the bin-opening during filling and to leave some material in the bin over the opening when refilling.

6.1.2. Belt Feeders

The belt feeder consists of a continuous rubber belt supported by closely spaced idlers and driven by end pulleys which are generally referred to as

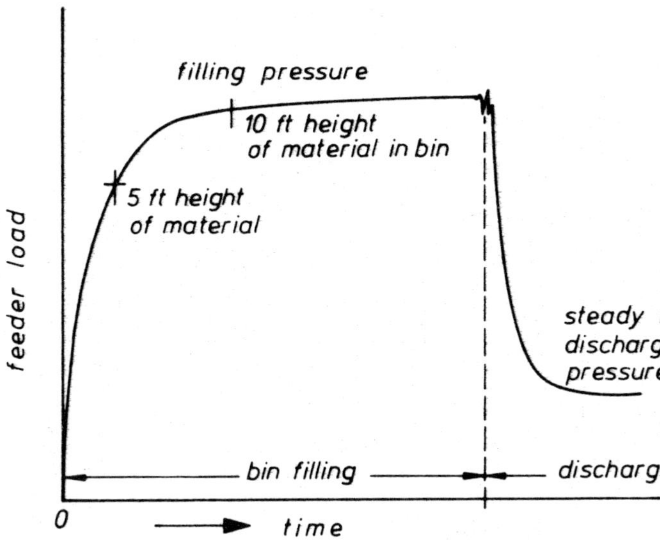

Fig. 96

the head and tail pulleys, see Fig. 97. This unit is contained within a single frame and the motor can be mounted on the ground or another frame and drive the feeder by means of V-belts. The belt feeder is usually placed under a long slotted hopper opening feeding along the length of the hopper. The slot-type belt feeder is one of the most economical feeders for bulk solids. Properly applied, the belt feeder lends itself nicely to low first cost, dependable operation and automatic control.

Belt feeders generally range in widths of from 2 to 6 ft and have lengths of 5 to 10 ft. The capacity of the belt feeder is dependent upon the width and rate of movement of the belt and is generally found to be between 5 and 1,500 tons per hour. The 1,500 tons per hour figure is a bit conservative. some belt feeders have been used for 2,000 to 2,500 tons per hour using

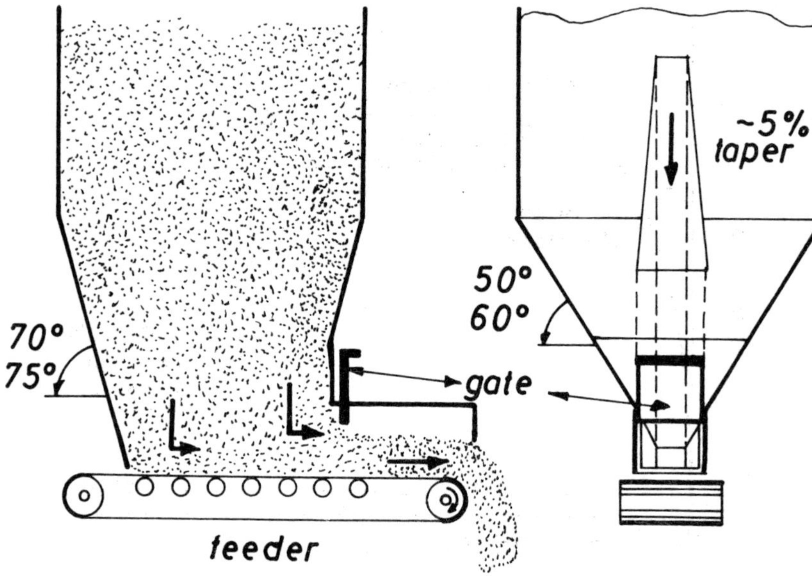

Fig. 97

special 8 ft wide belts. The rate of belt movement in feet per minute is one of the primary factors that determines the capacity of the feeder. Rate capabilities are from 1 to 100 feet per minute. Normally 50 feet per minute is considered a good belt speed when the belt is being used as a feeder. Higher rates cause excessive wear, lower rates require too large a belt. Power requirements for belt feeders vary between 3 and 50 horsepower with most of this horsepower being required for start-up due to the relatively high torques which are required to initially shear the material, particularly when the feeder is subjected to the high pressures under a bin which has just been filled (Fig. 96).

Belt feeders are used with fine ores, coal or concentrates, and occasionally will be used with very fine dry powders. They are well suited to the fine ores and also concentrates unless the concentrates become quite wet. When the moisture content of the concentrates exceeds 5 to 7 % the ores tend to cling to the belt. The same problem becomes more pronounced with fine powders and in either instance there are several considerations that have to be made. First, the belt should be supplied with a cleaning device on the return side of the belt which removes the fine material which is clinging to the belt. Second, hopper considerations become more important. When handling concentrates and fine powders, long slotted hopper openings are generally required to eliminate arching and ratholing; for the belt feeder to properly handle these materials and to eliminate flushing a double skirted arrangement is often advisable.

Belt feeders have several advantages. First, they are of relatively simple construction. They offer positive extraction of the material and fairly good feed regulation; they can be controlled by variable speed motors, either ac or dc. Belt feeders offer spillage free operation in the area of the hopper when properly skirted. One advantage of belt feeders is the relatively low power required for steady state operation. One objection to the belt feeder arrangements is the space required for the feeder, the height requirement is occasionally a problem. Space must be provided for the top surface of the belt, the return belt, the drive mechanism, the frame and the idlers. Belt wear is a problem particularly if the materials tend to be abrasive or if the particles are sharp or hot. In this regard the belt feeder should not be used with hot materials, or very large chunks or very abrasive materials. One very important thing is the hopper design which must permit material to move along the length of the hopper, if not, severe shearing conditions occur within the hopper opening and this tends to abrade away the belt much more rapidly than it would in an installation that has a good hopper-feeder combination.

The drive requirements for a belt feeder may be determined by considering all factors which tend to resist motion.

The belt pull is made up of the following different frictional and shear elements:

a) friction between belt and the slide plate or idlers,
b) friction between material and belt,
c) friction between material and hopper walls and skirting,
d) shear forces within the material at the gate regulating depth,
e) pulley and bearing friction of terminals.

Each of these elements should be calculated independently, but insufficient information exists today to evaluate them separately with good enough accuracy.

Further development work is required to establish more precise calculating methods for these elements. In lieu of a sound theoretical approach, the factors can be combined in an empirical relation which represents the effective belt pull (T_E) at the drive pulley during steady flow as:

$$T_E = \mu P_V \cdot B \cdot L + \Sigma F_x$$

μ = coeff. of internal friction in shear area

P_V = feeder load in psf

B = width of hopper opening in ft

L = length of hopper opening in ft

ΣF_x = sum of all frictional forces in the feeder

From earlier discussions it has been determined that P_V can be computed by using (c)γ B. The μ value has been established experimentally by the University of Hannover (113) and may be averaged as $\mu = 0.4$. Jenike suggests μ = 0.45. F_x may be lumped as $(0.1 \cdot \gamma \cdot L_c \cdot W^2 + 250)$ lb, assuming bed height to be about 1/3 of the belt width.

γ = bulk density in lb/ft^3
L_c = length of feeder in ft
W = belt width in ft

If we let (c) = 1.0 x), W = 2 B, L_c = 2 L and add additional pull to take care of starting conditions (accelerations, break-away, occasional larger feeder loads due to filling), then a good safe approximation for T_E is:

$$T_E = 2 \cdot \gamma \cdot L \cdot B^2 \qquad \text{(for slot opening)}$$

The feeder horsepower would then be:

$$h_p = \frac{2 \cdot \gamma \cdot B^2 \cdot L \cdot S}{33{,}000} \qquad \text{(for a slot opening)}$$

where S = belt speed in fpm.

For a square or round hopper outlet opening, we may use

$$T_E = 1.6 \cdot \gamma \cdot B^3$$

and

$$h = \frac{1.6 \cdot \gamma \cdot B^3 \cdot S}{33{,}000}$$

The above approximation is acceptable for most raw materials mined. In case of chemical powders more precise determination of P_v is required and flooding problems must be considered.

To maintain a uniform flow from the hopper opening, a slot feeder must have a taper in the direction of belt travel, as shown in Fig. 98. The exact

----

x) For most materials with $\delta = 40^\circ$ to 60°
$\phi' = 25^\circ$ and $\Theta' = 25^\circ$
the (c) values range from 0.8 to 1.0.

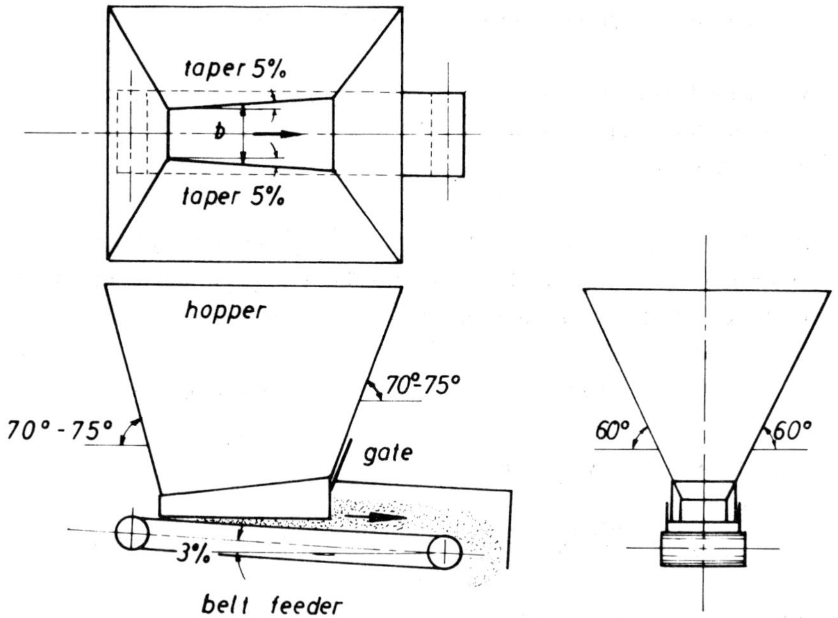

Fig. 98

amount of this taper has not been established yet, but in many installations a 4 to 5 % taper has worked satisfactorily. The taper also tends to reduce the normal running-torque by about 20 %. A slight decline of 3° of the feeder has also shown a beneficial effect on starting and running torques.

The bed height under the gate has a definite influence on the shear force (113). Beyond the restriction of clearance underneath the strike-off point being at least two times the maximum lump size, there is a width, length and height interaction affecting the shear force. It appears that the bed height should be at least 1/3 of the hopper opening width for slot ratios of L/B = 3. The bed height should increase for larger slot ratios in order to minimize the shear force.

The University of Hannover (113) has conducted a series of small model tests to determine the flow patterns in the hopper as a function of the gate position (bed height) on a belt feeder, as shown in Fig. 99 and 100.

In Fig. 99, the gate height is 25 mm above the belt. Only a small region of the stored material above the tail end of the feeder moves initially downwards. The front end of the bin discharges at a much slower rate, creating a considerable segregation. In Fig. 100, the gate position is 55 mm high providing a more regular motion of the material in the bin with less segregation. Because of practical considerations, the height of the gate position is limited and therefore, segregation cannot be completely avoided when using belt feeders.

6.1.3. Apron Feeders

The apron feeder consists primarily of chain linked heavy cast manganese pans, see Fig. 101. Usually a two strand chain supports the feeder pan on a center rail. For very wide feeders it is recommended that three strand chains be used. The rollers required beneath the feeder pan should be centrally lubricated, provide good tracking and proper support of the feeder pans. The hopper considerations for the apron feeder are in general the same as for the belt feeder. If it is to be used under a truck dump hopper with a long hopper opening, then the hopper should be tapered diverging in the direction of flow. It must be emphasised that apron feeders are used for high capacity large size materials and the hopper gate must be designed to permit very large chunks to come through the hopper opening. The feeder shown is provided with a method that overcomes potential hang ups. Chains are hung from the hopper outlet, or in this case caterpillar tracks are used, which aid in providing a flexible front wall for unrestricted flow of the large lump material.

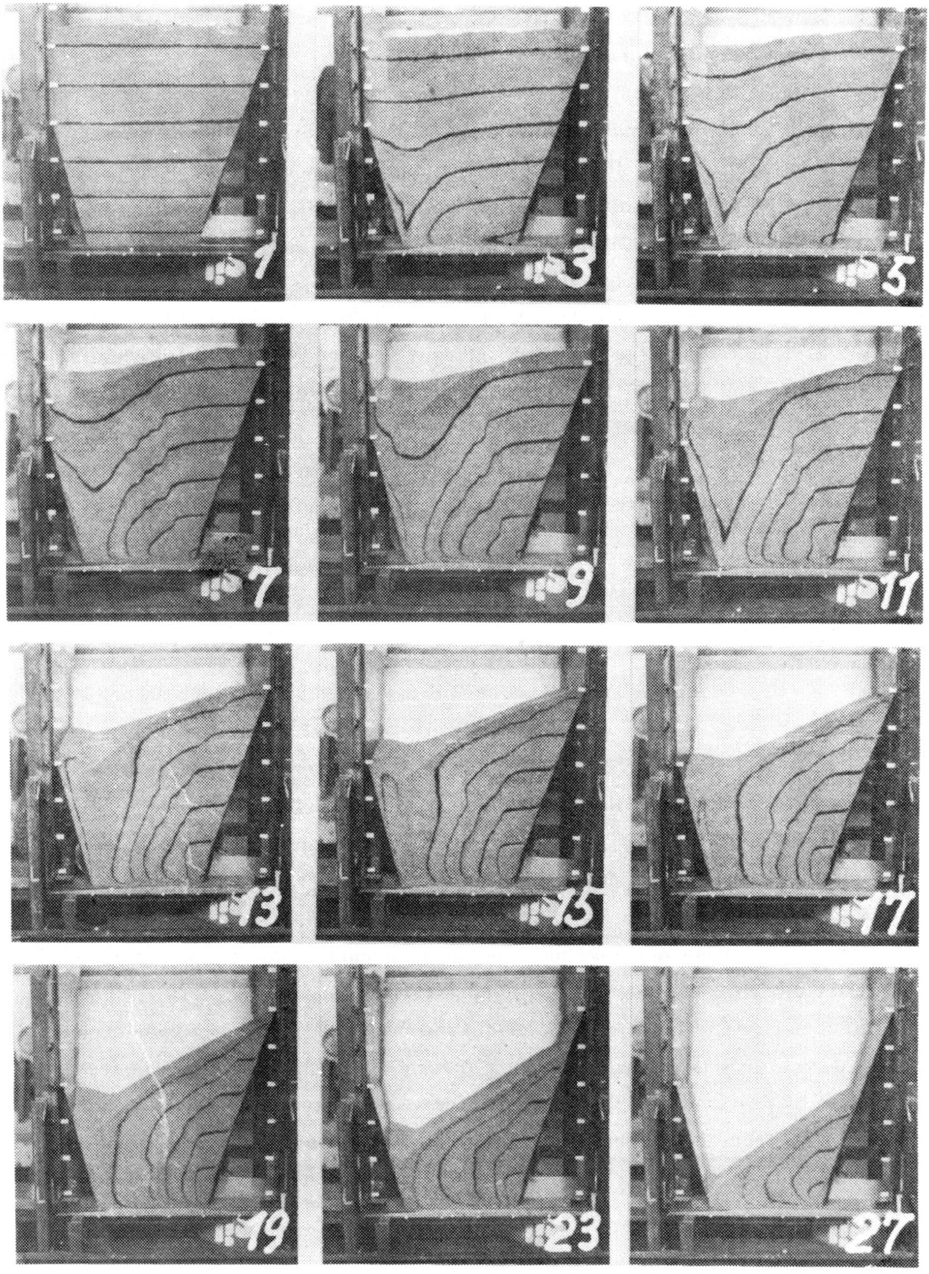

Fig. 99

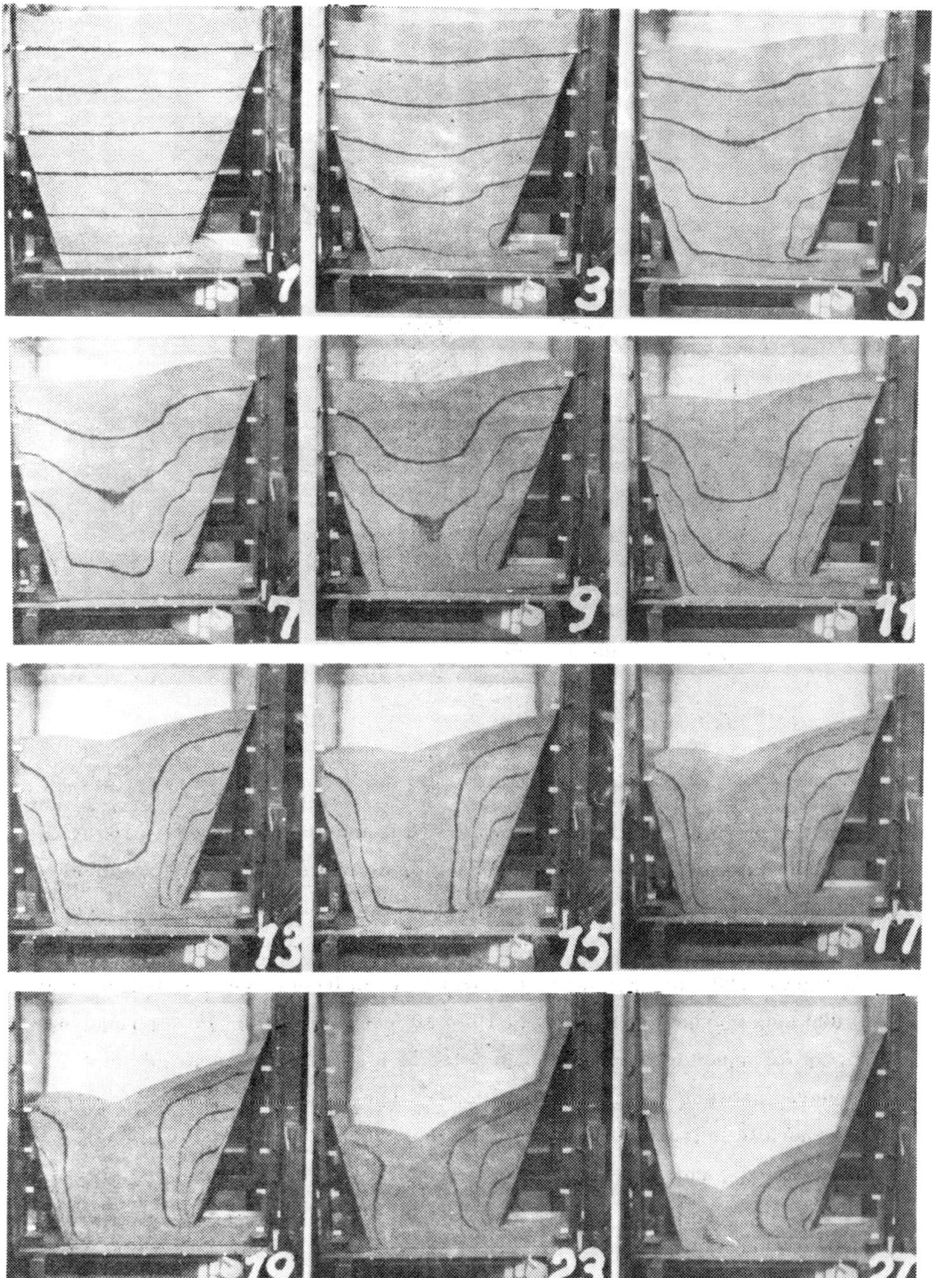

Fig. 100

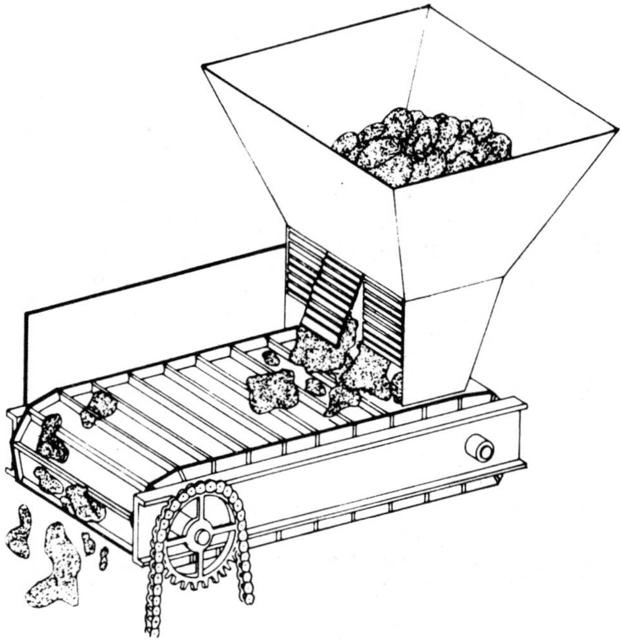

Fig. 101

Apron feeders vary in widths from 2 to 10 ft and lengths from 8 to 100 ft. The lengths in excess of 10 to 15 ft are used primarily for conveying material rather than as a part of the feeder itself. This is similar to the belt feeders which were discussed before. The capacities of the apron feeders are shown as 100 to 2,000 tons per hour. Generally apron feeders are limited to about 1,600 tons per hour with rates of 10 to 50 feet per minute. Power requirements for apron feeders are about twice as high as vor comparable belt feeders, between 10 and 150 horsepower. The application of apron feeders is generally in truck dumps or other areas where very coarse materials are handled such as feeding primary or secondary crushers. Other areas include car unloading facilities or quarries where the high impact resistance of the manganese pans is required.

The capacities of the apron feeder are dependent on the rate of movement of the pans, the density of material, the depth of flow and the width of the feeder itself. The width is between 2 to 10 ft as discussed before with the depth of flow generally not exceeding a half of the width of the feeder. The horsepower requirements for the apron feeder can be estimated by increasing the belt feeder requirements by a factor of at least 50 %. As with belt feeders the horsepower requirements listed by manufacturers are generally quite low and should be increased by a factor of 30 to 50 % to take into account considerations like starting torques, starting when cold, when the bearings are stidky, when the bearings become worn - all these things influence the power requirements.

The apron feeder has the ability to handle large tonnage capacities with positive feed extraction and uniform flow. It lends itself to speed control with the regulated speed ac or dc motors, and has one other advantage in that it can convey material uphill. It can handle large chunks and it is impact resistant which is very important in truck dump applications.

One disadvantage o the apron feeder is the space requirement. Apron feeders generally require a dribble chute beneath the feeder and also a belt feeder to remove the fine material from the dribble chute. The apron feeder does not offer the possibility of scalping and has a great many moving parts requiring a lubrication system and of course relatively high maintenance. Power requirements for the apron feeder, as discussed before, are about 50 % higher than those required for the belt feeders which for some applications may be considered high. One other disadvantage of it is the feeder requires a layer of fines on the pan if it is to be used under severe impact conditions.

6.1.3.1. Bar-Flight Feeder

Another type of high volume feeder for coarse materials is the bar-flight feeder, consisting of multiple chains interconnected by cross bars sliding over a steel base plate, Fig. 102 (92).

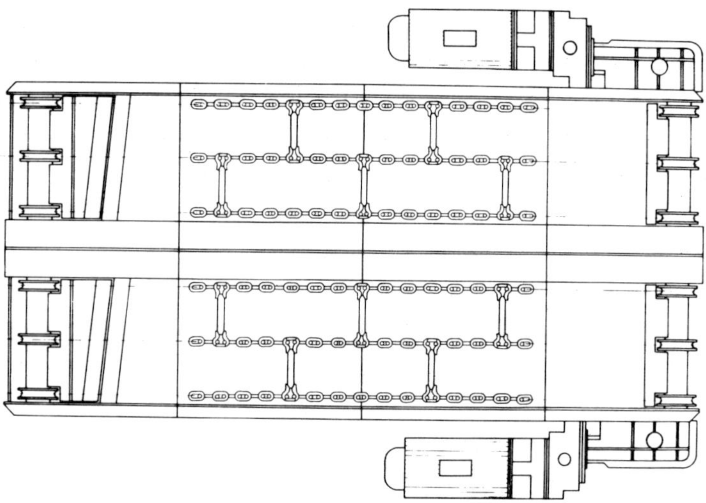

Fig. 102

The material is moved by the endless chains over the steel pan to the discharge point. Depending on the capacity requirements, the feeder may be equipped with two- to five-strands of chain or consists of parallel units (Fig. 102: two double-strand units). Generally, the maximum output is 4,000 tons per hour, but special constructions enable outputs from 5,000 to 10,000 tons per hour. Good experience with these feeders have been obtained in underground mining, in preparation and screening plants, in blast furnace stock houses, and garbage disposal plants (F 59).

The principle of an endless chain with scrapers can also be applied for the discharge of good flowing material from slot bunkers. Two strands of single chains pull the scraper along the shelf and discharge on a belt conveyor (Fig. 103) (122). A tent-shaped canopy is necessary over the whole length

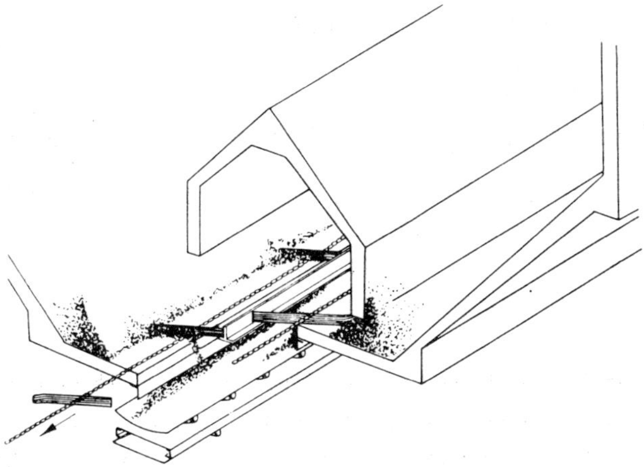

Fig. 103

of the tunnel for this arrangement. Good homogenization can be achieved as the material is discharged at many points at the same time.

6.1.3.2. Rotary Drum, Vane and Valve Feeders

A rotary drum feeder (Fig. 104 left) might be compared with a belt feeder with zero center-distance (181). The drum stops the material from freely flowing out, but it discharges by its rotation. The construction is cheap and easy to maintain, but it can only be applied for materials having a good flowability.

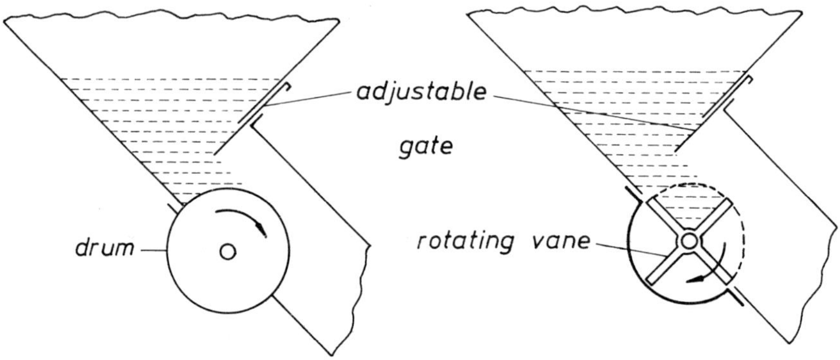

Fig. 104

Similar considerations apply to the rotary vane feeder (Fig. 104 right) which likewise might be compared with an apron feeder with zero distance between the centers (92). A modification of this type, using vanes with different pockets, is shown in Fig. 105. It provides an accurate feed at uniform

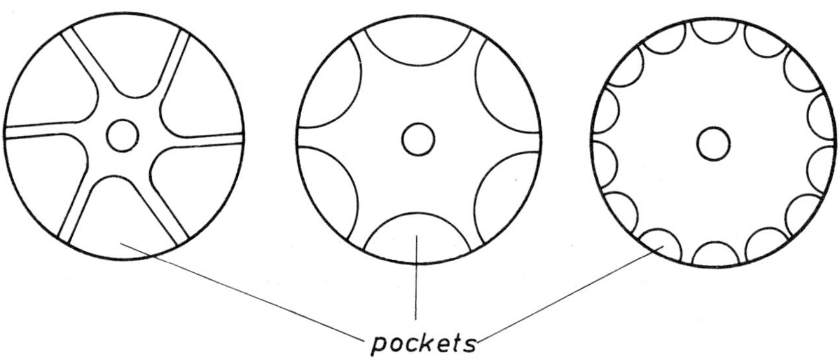

Fig. 105

density and is especially suited for feeding several materials to a mixing unit (F 13), (F 25), (F 33), (F 34).

The rotary valve feeder is completely enclosed and prevents powders or fine grained material from flooding (Fig. 106). The volume being discharged by rotation is given by the size of the pockets. Another advantage of this type of feeder is that it can discharge directly into a pneumatic conveying line (F 13), (F 33).

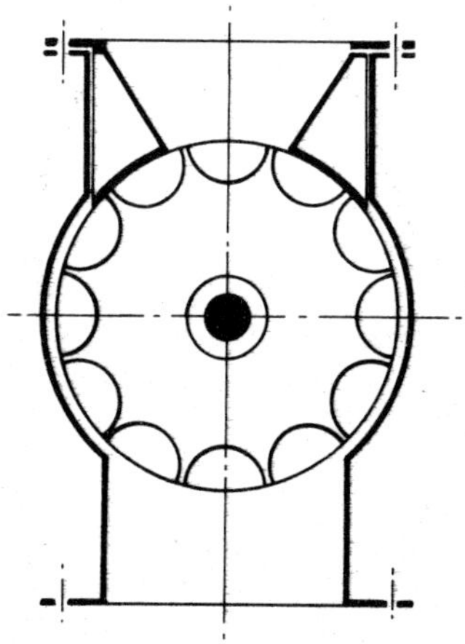

Fig. 106

6.1.4. Rotary Table Feeders

For sticky materials (wet ore fines) and relatively low feed rates (5 to 125 tons per hour), the rotary table feeder is often used (14), (198), (291). It can accomodate hopper openings up to 8 ft in diameter. The table-diameter is usually 50 to 60 % larger than the hopper outlet diameter. Rotating speed of the table ranges from 2 to 10 rpm. The drive horsepower varies greatly from one manufacturer to another. A lot depends on the proper configuration of the hopper outlet, outlet collar and plow position.

If the outlet collar is helical (F 34) or spiral as shown in Fig. 107, then fairly uniform flow can be expected in the hopper outlet. However, a dead conical mass will still remain on the center of the table, causing most of the shearing resistance. This mass occupies a cross-sectional area of about 40 to 50 % of the hopper outlet and is about half outlet diameter high.

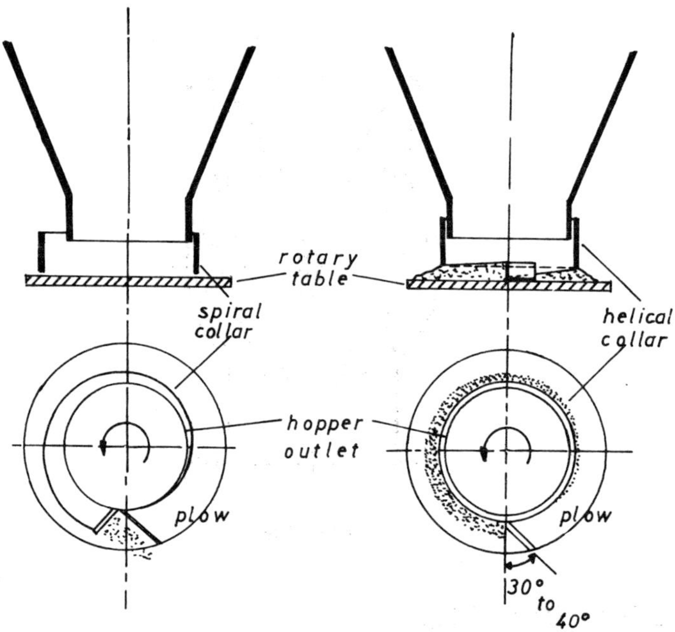

Fig. 107

A rotary table feeder consists primarily of a gear reducer and the cost therefore is greatly dependent on the torque required.

As in the case of belt feeders, little research effort has been done so far in determining the exact torque requirement of rotary table feeders as a function of material properties and hopper geometry.

Using the bin wall pressure (σ') as the feeder load the following approximation for the torque (T) under running flow condition may be used:

$$T = 0.13 \cdot (c) \cdot \gamma \cdot B^4$$

where

(c) = $\sigma' / \gamma \cdot B$

B = hopper outlet opening (ft)

γ = bulk density (lb/ft^3)

The horsepower to feed the material would be

$$hp = \frac{T \cdot N}{5{,}250}$$

where N = revolutions per minute

As the filling pressure on the feeder could be about two times larger than the steady flow pressure and further allowing for the drive losses, the required design torque for a table feeder becomes now:

$$T = 0.35 \cdot (c) \cdot \gamma \cdot B^4$$

and the required horsepower:

$$hp = \frac{0.35 \cdot (c) \cdot \gamma \cdot B^4 \cdot N}{5{,}250}$$

Rotary table feeders have their own surge hopper. They can handle sticky materials and are relatively simple in construction therefore requiring very little maintenance. However, some particularly cohesive materials will tend to cling to the hopper walls and may skid on the surface of the table and not be removed at all. Other things that may be objectionable are the high installation cost of the rotary table feeder and the mammoth piece of machinery that is required for low tonnage output.

6.1.4.1. Circular Bin Dischargers

This type of dischargers has been developed for use with circular bins. It provides a completely enclosed, dust-free method of discharging powdered, granular or small lumpy materials which will not readily gravitate from bins. A central cone fitted with protruding rods revolves slowly and sweeps over the entire base of the bin, discharging the material through one or more outlets (Fig. 108). Volumetric feed control of considerable accuracy is possible for many materials. An arch-breaker (see Chapter 5.2.3.) can be attached through a universal joint to the top of the cone, so that it will move with the cone (181), (261), (317), (F 49), (F 55). The range of capacity for this type of circular bin discharger is given in Table 12 (F 2):

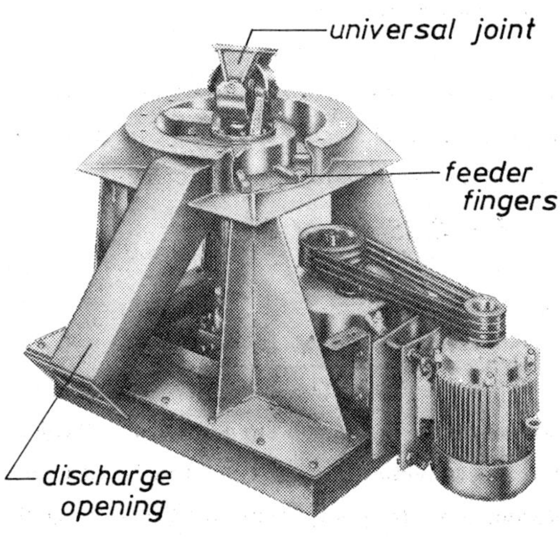

Fig. 108

Table 12

size of bin discharger (in.)	18	30	48	60
discharge volume (cu. ft. /rev.)	0.264	1.34	6.14	10.6
maximum rotor-rpm	16.0	10.1	6.4	5.2
discharge volume at max. rpm (cu. ft. /min)[x)]	4.22	13.6	39.4	55.0
max. lump size[xx)] (in.)	1	2	$2\frac{1}{2}$	$2\frac{1}{2}$

x) based on one discharge point

xx) fragile materials only

A somewhat different design consists of a fixed circular shelf and one rotating plow (Fig. 109) (F 8). The power requirements of rotating the plow are less than 25 % that of a comparable rotary table feeder. Bin bottom openings could be as large as 25 ft in diameter. The cone in the bin diverts the material to the circular shelf, where it is plowed off. This results in a mass flow pattern inside the bin. A wide range of discharge rates (up to 500 tons per hour) can be obtained by varying the speed of the gathering arm. These installations have been constructed for sticky materials, such as wet limestone fines, ore concentrates, gypsum fines, wet lignite and coal fines. They can be completely emptied and even large lumps of 4 to 12 inch in diameter can be handled provided they are mixed with finer material. For example, a size consist of 70 % minus 4 inch and 30 % 4 to 12 inch material presents no problem (1), (160).

Another type of circular bin discharger is a combination of two rotating arms and a fixed screw feeder (F 52). It is specially designed for the storage of wood chips and saw dust up to very high moisture contents. The rotor unit loosens the bulk material in the lower part of the bin, and feeds it continuously to the screws which are located at the bottom, from the

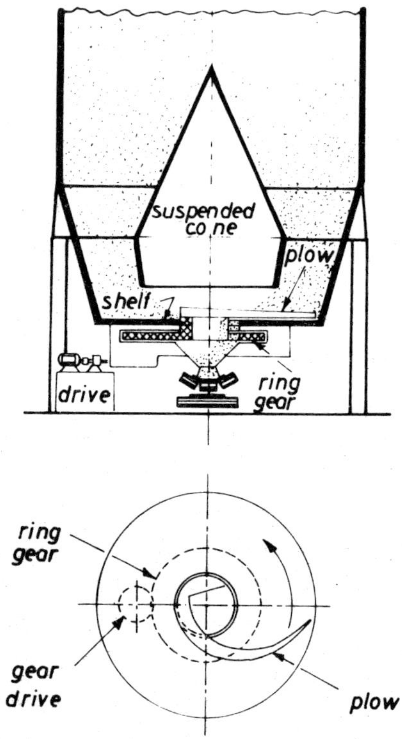

Fig. 109

axis of the bin through the wall. The screws are tapered and have a progressive pitch. The number of screws depends on the material and the discharge rate. Every screw section represents an integral unit of trough, drive and mechanical control.

6.1.4.2. Rotary Plow Feeders

This type of feeder has not made as much inroad in the USA as it has in Europe. It was first developed in Germany about 40 years ago for feeding of lignite (F 8), (F 37). Since then it has found a wide field of application for other materials, such as sinter, coal, potash, phosphate, limestone, iron ore, cement clinker, etc. Rotary plow feeders are suitable for use in tunnels, under storage piles, or under long storage bins.

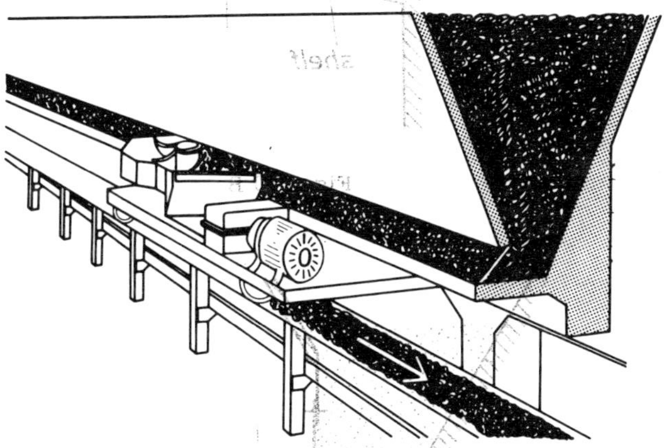

Fig. 110 A

Fig. 110 shows the basic principles of the rotary plow feeder. The plowing mechanism consists of curved armes arranged to sweep material off a narrow shelf running the length of the storage pile or bin, Fig. 110 A (161), (253). Different cross-sections of the outlet and the method of operation of the plow are shown in Fig. 110 B and C.

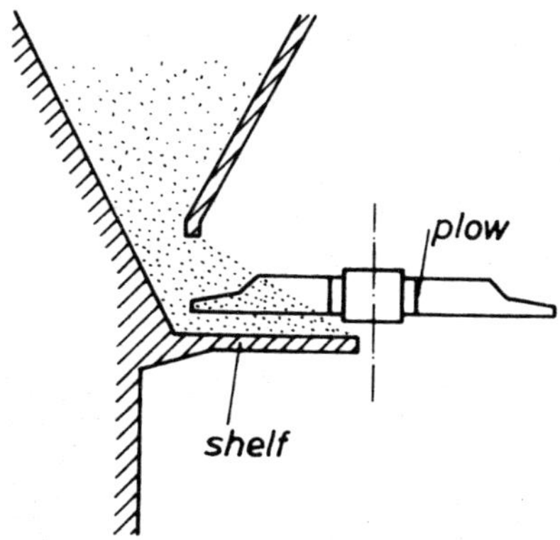

Fig. 110 B

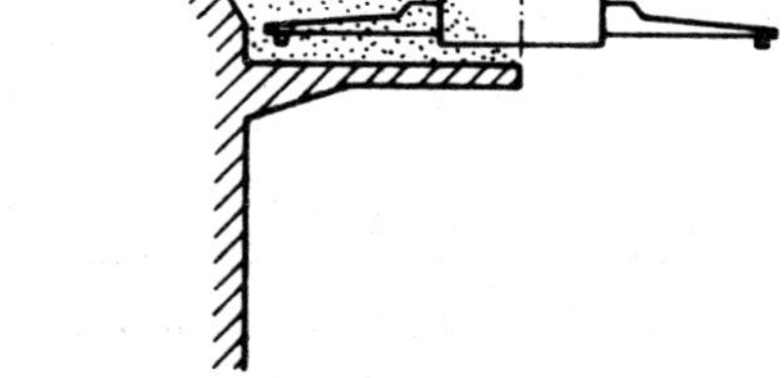

Fig. 110 C

B: The pressure of the bulk material in the bin only acts on the inclined hopper walls. The plow removes that volume which is resting on the shelf. This method may be successfully applied with good flowing but non-flushing materials, like dry sand, classified coal etc.

C: In this case the hopper has a shape which allows the pressure of the material to act directly on the shelf. The plow arms are not as high as in the other picture (B) but they dislodge the material in the pressure zone. The chance of forming stable arches is considerably reduced. This method of discharge may be recommended for poorly flowing materials, such as wet clay, sticky ores etc.

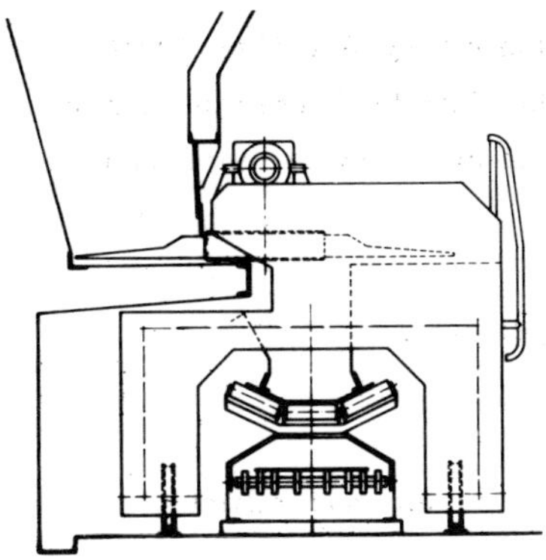

Fig. 111 A

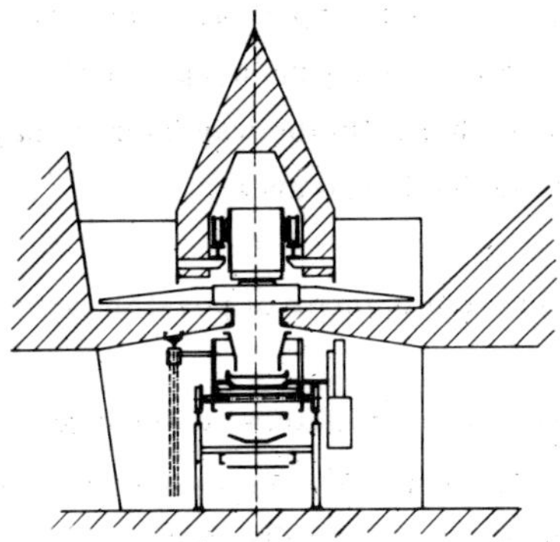

Fig. 111 B

Fig. 111 A and B are schematic cross sections of different installations. In 111 A the plow discharges from one shelf unilateral on a belt conveyor. The plow unit moves on rails along the bin. In 111 B the plow discharges simultaneously from two shelves bilateral on a weigh belt, through which the rate of flow resp. the rotation speed of the plow is automatically controlled.

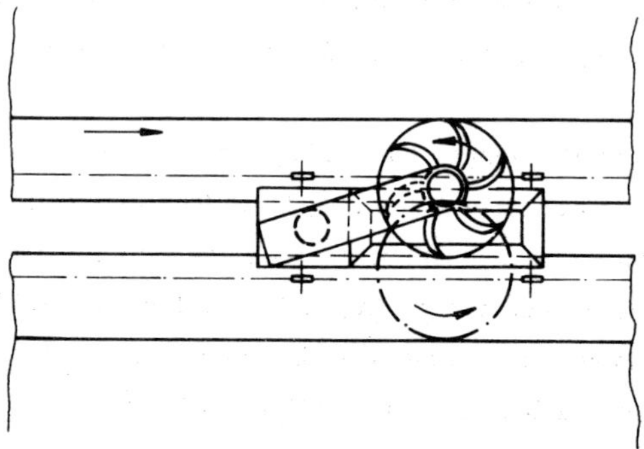

Fig. 111 C

Another method is the application of a slewing rotary plow (Fig. 111 C) (161). Travelling along the hopper outlet, the plow discharges the material from the one shelf. On the way back it may be moved to the other shelf, enlarging the discharge area.

Fig. 112 shows a number of possible arrangements:

a) standard shape, unilateral discharge,
b) two plows, larger discharge area and volume, for most difficult materials, unilateral discharge,
c) two plows, discharge unilateral under a tent-shaped roof, discharge area even more enlarged,
d) one plow, discharging bilateral; more of these installation may be combined so that relatively large bins or storage piles can be discharged,
e) for the discharge of flushing material this special arrangement uses inclined shelves and correspondingly inclined plows (F 28).

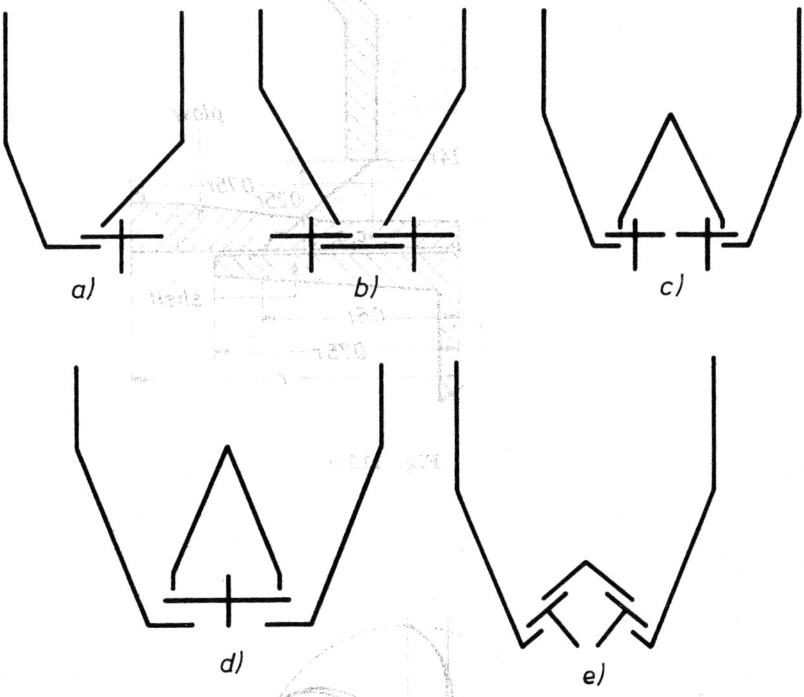

Fig. 112

Although the hopper opening, shelf width and height will depend greatly on the type of material handled, the basic dimensions may be generalised for materials with δ = 35 to 50 degree, and expressed in terms of the plow radius R (see Fig. 113 a and b).

Assuming two blades active and operating on one shelf, the horsepower equation for the plow drive may be approximated by:

$$hp = \frac{T \cdot N}{5,250}$$

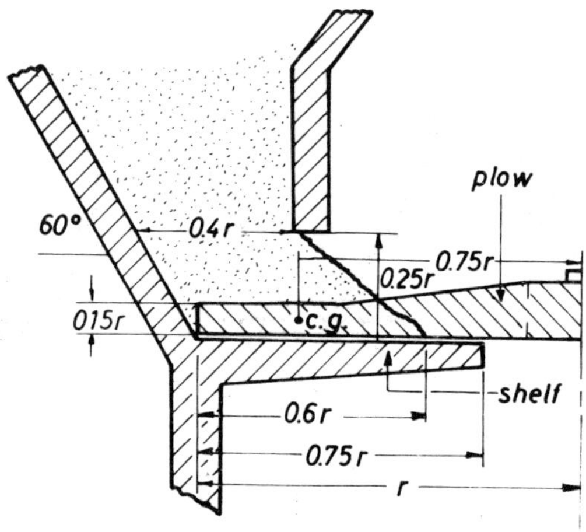

Fig. 113 a

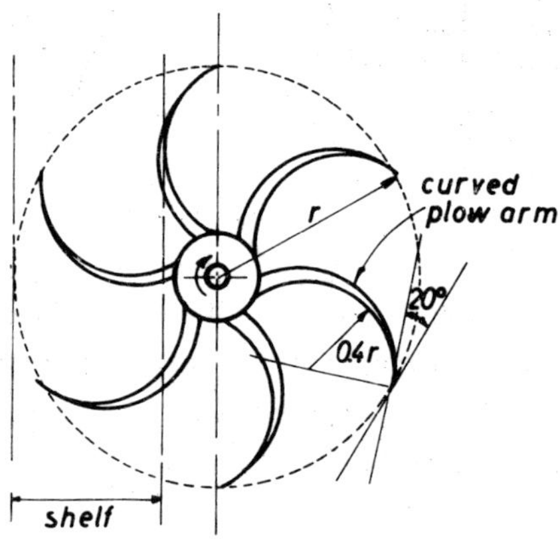

Fig. 113 b

Fig. 113 c

where

$T = 0.15\ R^4$ ft-lb (torque)

N = rpm

γ = bulk density (lb/ft^3)

R = radius of plow (ft)

The feeder capacity is approximately: $t/h = 0.01 \cdot \gamma \cdot R^3 \cdot N$.

The rotating speed N is usually kept below $350/\pi \cdot D$, where 350 feet per minute is the maximum suggested tip speed of the plow.

The proper curvature of the plow arm is important to obtain the best efficiency. For good flowing materials they usually have a radial curvature. For difficult materials a special curvature as shown in Fig. 113 c works very successfully (F 37). Usually the plow consists of four to six arms with a diameter of 6 to 16 ft. Because of the relatively large number of moving parts, the maintenance could be a problem. But since the plow arms can be examined at every rotation this problem is reduced considerably.

The discharge rate ranges up to 3, 500 t/h, and is mainly influenced by the size and operation of the feeder. On the other hand, it also depends on the flow characteristics of the bulk material. The following table gives an example of discharge rates for various plow diameters for classified coal using a unilateral discharge (F 37):

plow dia.(mm):	1, 600	2, 000	2, 500	3, 000	4, 000
capacity (m^3/h):	150	350	700	1, 000	1, 500

6. 1. 5. Screw Feeders

A screw feeder (Fig. 114) is essentially designed for very low tonnage outputs where the assurance of positive discharge is a must and it offers an advantage in that the feeder itself can be easily enclosed making it dust tight. Thus it provides a closed hopper and chute arrangement from the hopper to delivery point. The feeder consists primarily of the helical screw rotating beneath the hopper outlet driven from an external source. The screw itself can be of a fixed pitch or can be of a varied pitch having a smaller pitch spacing in the rear gradually increasing in pitch to the discharge end. This ensures that the material will be moving in the back portion of the hopper. Occasionally screw feeders will be required to have tapered screws, that is a smaller diameter in the back, gradually increasing to the largest diameter at the outlet. This taper assures near uniform material removal from the hopper outlet.

Different types of screws are shown in Fig. 115 (F 11), (F 21), (F 24), (F 26), (F 34), (F 55).

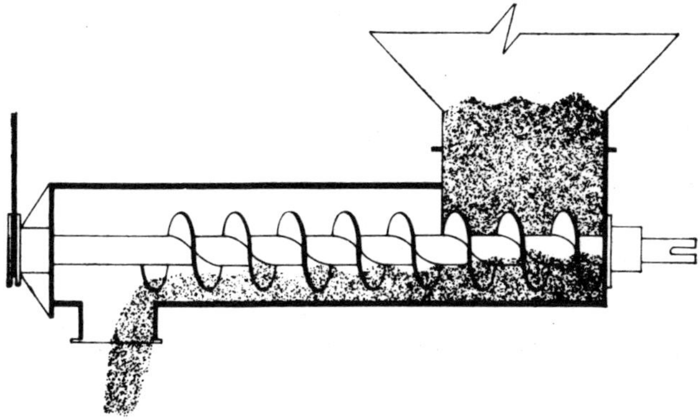

Fig. 114

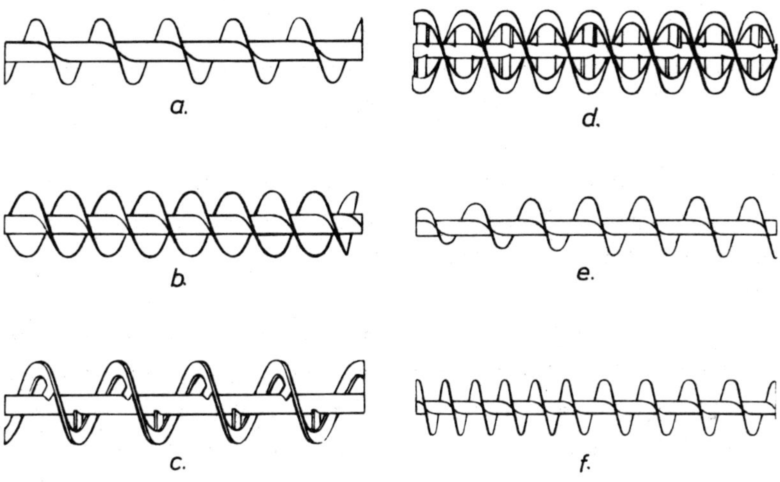

Fig. 115

a) Standard-pitch: It assures smooth performance. The flight is welded to shaft creating strong one-piece construction. Various standards of thickness and welding are available to suit wear.

b) Double-flight: It delivers a more uniform stream of material than a single flight, and provides a smooth discharge where a more accurate control of capacity is required.

c) Ribbon-flight: It is adapted to handling of sticky or gummy material, which normally collects on the flights of standard screws where they join the shaft. The open space between shaft and flight eliminates the difficulty.

d) Double-ribbon flight: It is used to handle sticky or gummy materials. Double ribbon delivers a more uniform stream of material and provides a more even discharge.

e) Taper-screw: It is mainly used in feeders and hopper dischargers. The progressive increase in conveying capacity avoids dead pockets in a storage bin by extracting material along the full length of the exposed screw.

f) Variable-pitch: It is used to prevent a long feeder being overloaded when the feed-in point is "drowned" with material. It also prevents aerated materials "flooding" out of bins.

The entrainment pattern of the stored material in the screw is the feature which determines the pattern of flow across the hopper outlet slot. Where no inflow may take place there will be a "dead" region in the foregoing space (Fig. 116). Such a region does not allow a mass flow and may cause deterioration of the flow properties of the static material and all other consequent disadvantages (18), (29), (31), (33), (51), (147), (248), (315). Typical flow patterns presented for various screw forms are shown in Fig. 116:

a)

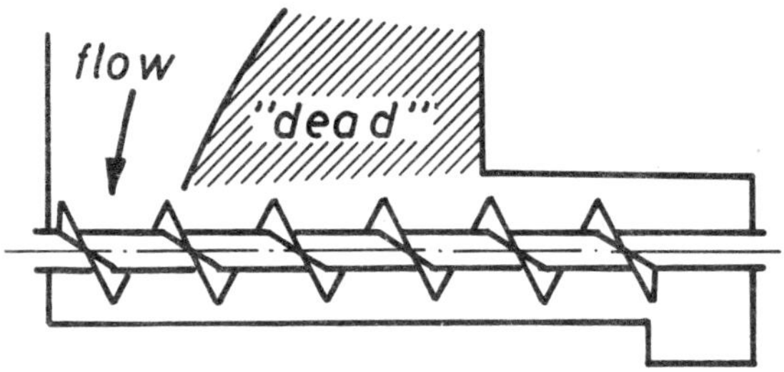

b)

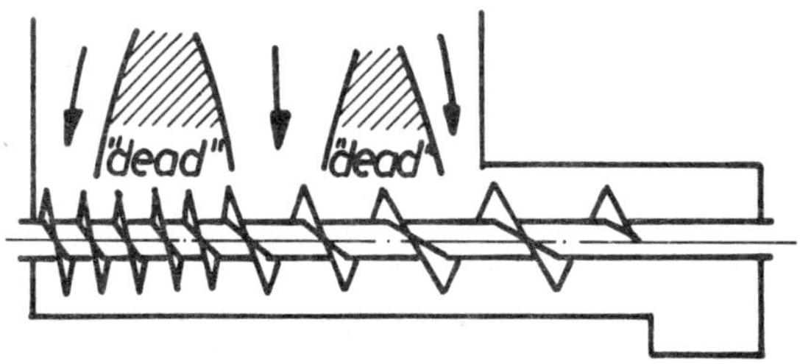

c)

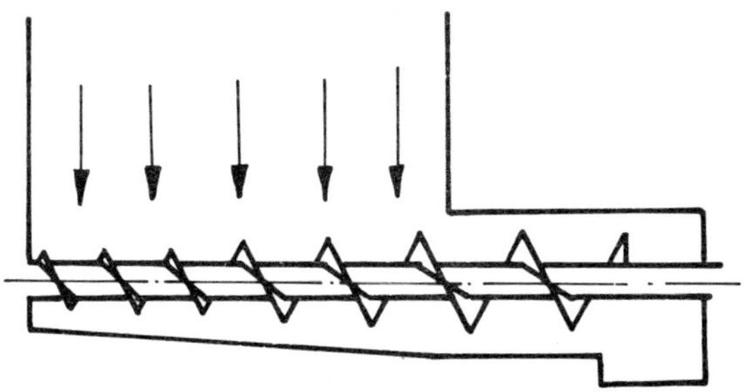

Fig. 116

a) uniform pitch and uniform diameter,

b) graduated pitch and even diameter,

c) increasing pitch and increasing diameter.

Screw feeders are limited in size from 2 to 24 in. in diameter and have lengths of from 2 to 20 ft. Capacity of the screw feeder is relatively low between 2 and 200 tons per hour. It is generally applied to very fine materials and powders giving flow rates between 2 and 40 feet per minute, depending upon the rotating speed and pitch of the screw. The horsepower requirement is between 1 and 10 hp. It can be applied to abrasive materials, however special considerations are required for loading and plow speeds must be limited since the materials are in contact with the screw sliding throughout the length of the feeder and cause considerable abrasive wear.

For dimensioning a screw feeder an empirical equation may be used which incorporates the length of the feeder, the hopper length, the diameter of the screw, and the density of the material.

$$hp = \frac{C(L_C + 3L)[(1.85 - 0.45D)\gamma + 19 - 8.85D]}{500}$$

C = capacity	(t/h)
L_c = feeder length	(ft)
L = hopper length	(ft)
D = screw diameter	(ft)
γ = density	(lb/ft^3)

Screw feeders can be used where there is a low head room requirement. They are good for handling fine powders, have a relatively low power requirement for the delivery rate, assure a positive movement of the material and offer a simple dust tight construction. The disadvantages of the screw feeder are: it has a very limited capacity and cannot handle large or abrasive materials.

Besides the conventional application of a single screw, several screws parallel to each other may be used for a larger outlet and to increase the discharge rate (147), (261). Another kind of screw feeder is applied for discharging a long slotted bin. The screw with the drive is supported on both ends on rails and traverses along the whole length of the bin bottom, discharging on a belt conveyor (Fig. 117) (148).

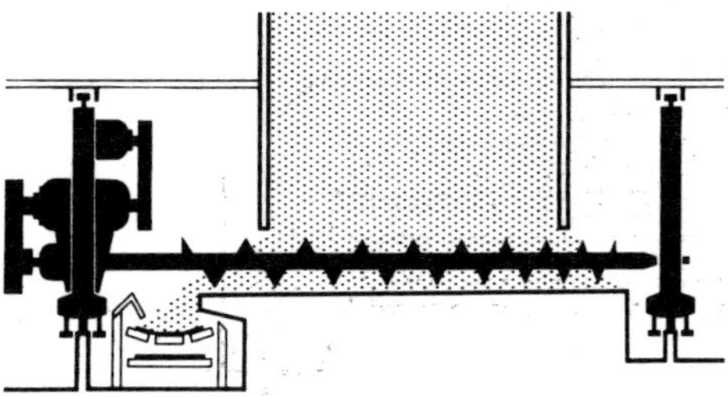

Fig. 117

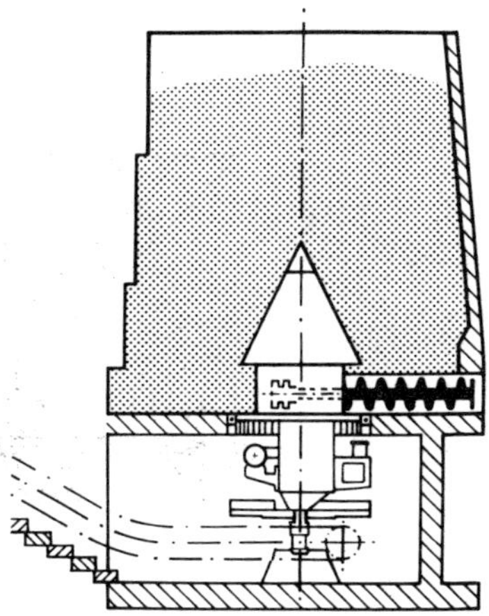

Fig. 118

Specially designed for the storage of clay, Fig. 118 shows a screw-type discharger circling slowly around the axis of the silo. Through the rotation of the screw the material is cut away and discharged into a collecting plate (F 26). The pressure exerted by the material on the screw can be reduced by the cone shaped cover on the rotating hub.

6.1.6. Reciprocating Plate Feeders

The reciprocating plate feeder (Fig. 119) consists of a horizontal or slightly inclined steel pan which usually rests on rolls or is suspended from chains. By means of a crank mechanism the pan is subjected to a forward and backward motion. During the forward stroke, the material is discharged from the hopper into the pan. During the backward stroke, the material is pushed off the pan by the back plate of the loading chute. The rate of flow is

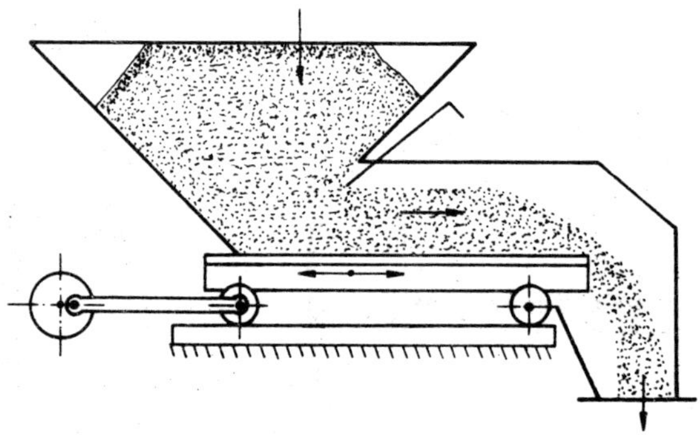

Fig. 119

$$\dot{M} = W \cdot h \cdot s \cdot n \cdot \varphi \cdot \gamma \cdot 60$$

where W = feeder width (ft)

h = vertical distance between pan and bottom edge of the gate (ft)

s = stroke of the pan (2 to 7 in)

n = frequency (20 to 60 strokes per min)

φ = coefficient of loading (0, 65 to 0, 70)

γ = bulk density (lb/ft^3)

The flow rate may be controlled by moving the gate or by variation of the stroke.

Plate feeders are generally used for the discharge of fine and medium materials. Because of their simple construction and low costs, they are widely used (219), (279) in the coal mining industry.

6.1.7. Vibratory Feeders

The process involved in determining the design parameters of a vibratory feeder is rather complex. Many papers on this subject have been published during the last ten years. Some examples are (17), (24), (67), (71), (97), (110), (112), (120), (124), (204), (216), (235), (239), (260), (266). Material on the trough is subjected to the forces of gravity, normal forces, friction forces, and impact forces. Basically the feeder trough or pan is driven by a nearly sinusoidal force at some angle to the trough. When the feeder is operating the trough is oscillating along a straight line, the amplitude and direction determined by the driving force.

The resultant linear vibration is a repetitive series of throws and catches that moves the material on the trough, Fig. 120 demonstrates the action for a single particle on the trough. The particle is in contact with the trough for approximately one fourth of the drive cycle shown as point a to point b. When the particle leaves the trough it travels with a uniform horizontal velocity but the vertical velocity gradually decreases due to gravity. At some later point the trough again contacts the particle and the process is repeated. This process then conveys material along the trough from 0 to 80 feet per minute depending upon the combination of drive frequency, amplitude, drive angle b and feeder inclination. These parameters as well as material flow depths and feeder trough widths allow material to

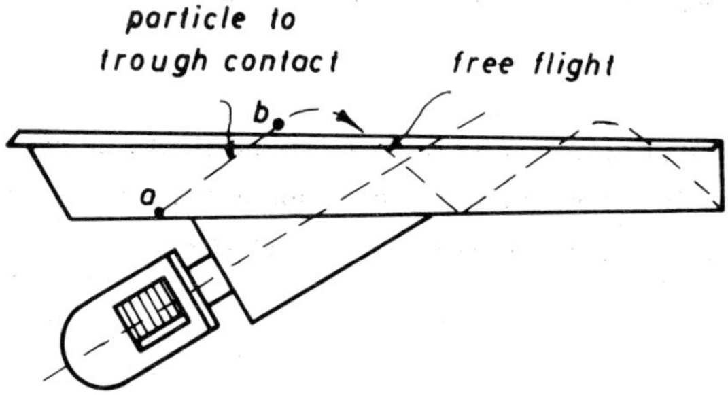

Fig. 120

be delivered at rates from several pounds to more than 3,000 tons per hour.

Various manufacturers of vibratory feeders have selected different operating parameters for the trough movement. Operating frequencies generally vary from 600 to 7,200 vibrations per minute, amplitudes range from a few thousandths of an inch up to one inch or more, and selection of drive angle b goes from 20° to 45°. For any given material there will be an optimum operating condition combining frequency, stroke and drive angle; it would seem reasonable for example that for particles such as fine limestone, flour and cement that tend to cling to the trough, more of a vertical component of stroke should give higher delivery rates. At the same time, abrasive material such as scrap steel or coke which flow easily would do less damage to the conveying surface of a short stroke, high frequency machine. However, selection of stroke, drive angle and frequency is not always sufficient to ensure a material's conveyability. For some materials, composed of very fine particles, additional modifications must be made to the feeder troughs, such as roughing the trough surface, or adding air to avoid a partial vacuum formed when tightly packed fine materials are fed along the pan. A list of these modifications is given by Guthrie and Morris (112).

For a given drive angle the delivery rate depends approximately on the product of frequency and stroke, so that in lowering one the other must adjust for this. For short stroke, high frequency (3, 600 cpm) operation electromagnetic units are chosen, conversely for long stroke, low frequency operation various choices of motors with eccentric weights are being used.

Generally the magnetic feeders operate at 3, 600 vibrations per minute at 60 Hz (line frequency), 3, 000 vpm at 50 Hz (line frequency) whereas the electromechanical feeders operate at 1, 800, 1, 200 or 900 cycles per minute at 60 Hz (line frequency), 1, 500 or 1, 000 cycles per minute at 50 Hz(line frequency). The range of operation for the 3, 600 cycle per minute feeders is between 35 and 65 to 70 mils stroke, at 3, 000 cycles per minute the stroke may approach 80 mils. Only recent advancements in the feeder industry have permitted the operation to be in excess of 65 or 70 mils stroke. In this region the feed rates get relatively high.

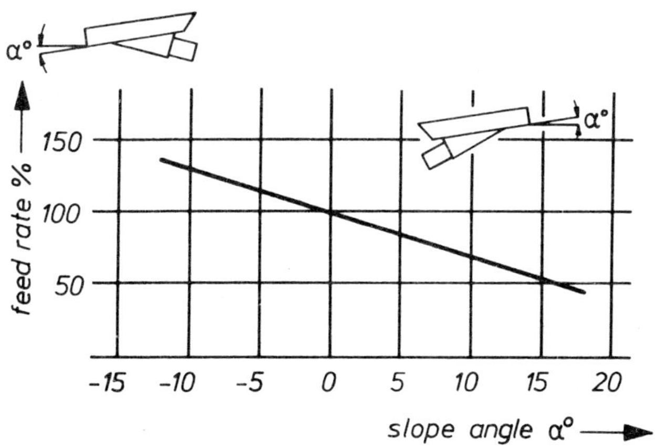

Fig. 121

Increasing the downward slope of the conveyor equipment will increase flow rate and throughput. As shown in Fig. 121, each degree of change of the angle of slope, between - 12 and + 12°, will alter the flow rate by approximately 3 % (F 1). Depending on the site conditions, increasing the slope angle up to about 12° may be advisable. It is not always possible to make use of these possibilities in combination. In practice, it is advisable to keep some reserve capacity in hand. The actual conditions on site are often difficult to assess. Furthermore, all conveyable materials vary to some extent in their properties.

6.1.7.1. Design Types

The drive systems for vibratory feeders can be categorised as either "brute force" or tuned systems depending on how the force is applied to the trough. Brute force machines may be considered where first cost is of prime importance, but it should be recognised that operating costs will be higher and greater forces have to be accommodated within the design. Conversely, the electromagnetic two mass subresonant tuned system has the lowest operating cost and requires virtually no maintenance. Other considerations in choosing the right design for a particular application include trough weight and dimensions, environment, the type of material being handled, and of course the feed requirements for the process. Fig. 122 and 123 show a comparison of the two types.

Generally the brute force or "direct force" feeders operate by having rotary vibrator drives attached directly to the trough. The drive will be through the center of gravity of the trough as shown in the sketch of Fig. 122 a. The vibrator drives are nothing more than eccentric weights on a rotating shaft. Some brute-force drives have dual motors that will run in synchronism because of a natural phasing phenomenon, Fig. 122 b. Other types have a single motor driving a dual set of gear connected weights through a V-belt, or will have the single motor pivoted as in Fig. 122 c to allow transmittal of linear force only and allowing adjustment of the drive angle. Feeders with

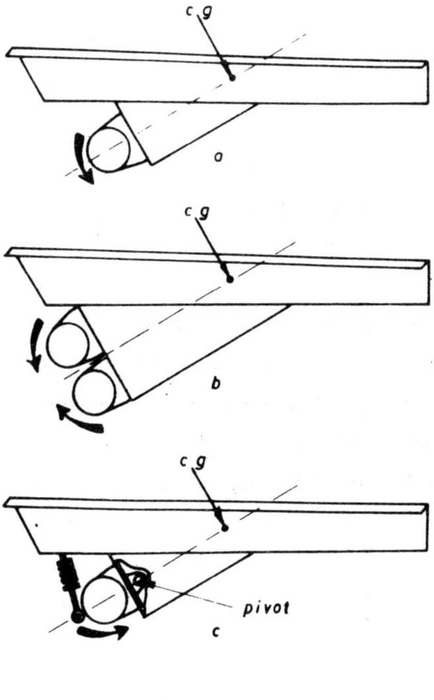

Fig. 122

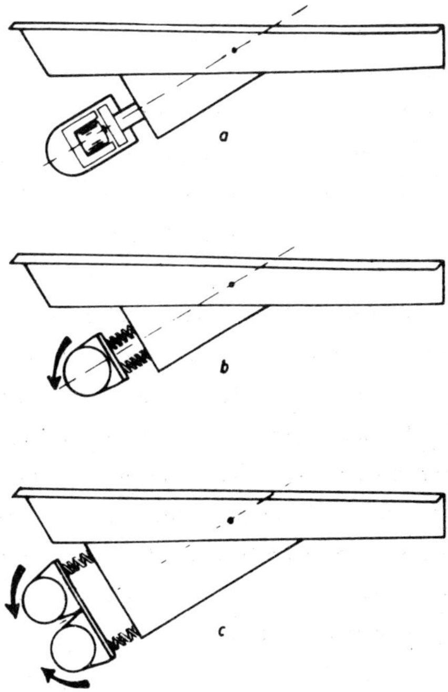

Fig. 123

a single set of eccentric weights as in Fig. 122 a are for light duty applications. The single pivoted motor of Fig. 122 b is for heavier applications and long conveying, while the systems with dual motors or dual sets of geared weights are for heavy duty applications - feeding primary rock crushers or conveying over relatively long distances.

The brute force feeder is used primarily where a single feed rate is required. Because of its design, control of flow rate is limited to changes in gate opening, or mechanically changing the feeder slope or value of the eccentric weight. Some manufacturers supply a variable speed drive to control flow rate. The advantage of the brute force system is that tuning need not be considered, that is, any drive frequency is acceptable.

One disadvantage of the brute force feeders is that the impact load of materials on the pan is felt directly by the bearings on which the weights rotate. In addition to this there are high motor loads on start up, requiring large horsepower drives for a given application. The delivery rates of the brute force machines are dependent on head load, a point that limits their application where uniform flow rates are a process requirement.

The second type of design is called a "tuned" or indirect force design. This is by far the most popular type of vibratory feeder. Fig. 123 depicts several of these designs. The tuned feeder is essentially a two mass, spring connected vibratory system. One of the masses is the trough, the other mass being the reaction mass. Because of the selection of spring constants a relatively small exciting force is amplified to generate a vibratory motion. In essence a tuned feeder takes advantage of the natural magnification of vibrational amplitudes that occurs when a vibrating system is operated near its resonance condition.

Drive force may be supplied by either electromagnets or rotating eccentric weights.

The capacity for vibratory feeders is dependent upon the width of trough, the depth of flow, density of material and, of course, the rate of material flow which is a function of the material itself. Depending upon the material size, density, particle configuration, and moisture content, maximum flow rates can vary from 10 up to as high as 60 feet per minute and more. Most common materials can be considered to be in the 30 to 50 feet per minute category. This includes materials such as coal, iron ores, limestone, and rock. Manufacturers of the vibratory equipment generally publish information on the rates of flow that can be expected. The depth of flow on the feeder generally should not exceed one half the width of the trough; however, for materials which are light and of coarse to medium size such as run of mine coal, the depth may exceed this one half width figure. Once the width of the trough is determined based on capacity, the length is calculated, based on the hopper opening and the angle of repose of material. The horsepower can then be estimated for tuned feeders, using a factor of 0. 15 times the area of the trough in square feet. The brute force machine, however, requires about 0. 2 as the multiplier for the area in square feet. These factors, as a rule of thumb, give good estimated horsepowers if the feeder size is known.

$$C = \frac{W \cdot d \cdot \gamma \cdot (\text{rate of flow})}{33.3}$$

rate of flow = 30 - 60 fpm

$$d = \frac{1}{2} W$$

tuned: $hp = 0.15 \cdot L_c \cdot W$

brute force:

$$hp = 0.2 \cdot L_c \cdot W$$

where:

C = capacity		(t/h)
γ = density		(lb/cu ft)
d = depth of material on feeder		(ft)
L_c = feeder length		(ft)
W = feeder width		(ft)

The power required to deliver a given capacity is dependent on several factors: the type of feeder, the type of material (its moisture content, density and cohesiveness), the hopper configuration and the required location of material delivery. All of these factors combine to produce a feeder trough of a particular size and weight. The trough weight, more than any other single factor, determines the drive size requirement. For a given capacity requirement it is best to use as short a trough as possible, thus minimizing the trough weight and consequently the drive size required.

The various vibratory feeder manufacturers have different trough designs and for tuned feeders different ratios of trough side to drive side weight which makes it difficult to assign even an empirical equation to define power requirements for a given capacity. However, for a "ball park" figure a rule of thumb would require at least one horsepower per hundred tons capacity on properly sized brute force machines. For two-mass tuned systems about one-half horsepower per hundred ton capacity is sufficient. The above estimates are based on a free flowing material with a density of 100 pounds per cubic foot.

The brute force feeders offer high capacity, the ability to handle large chunks, and have excellent impact resistance. Installation cost is low and because of the few moving parts it offers low maintenance costs. They have disadvantages, however; feed extraction is not fully positive and it has limited speed variation which is perhaps its biggest deficit. Fines may build up on the pan, particularly very wet concentrates. Stocky material may cause a problem having feed rates somewhat lower than would normally be expected.

The tuned feeder offers a very simple construction, high capacity feed; it is adaptable to automatic control systems because most of the control circuitry is of an electrical-electronic nature and can be tied in easily with feedback systems. Like the brute force there is no spillage in the area of the hopper. It gives fairly uniform flow and is generally mounted on a resilient shock absorbing suspension. It offers the possibility of scalping, a grizzly can be used on the front end of the trough. For magnetic feeders, virtually no maintenance is required - there is no lubrication as there are no sliding or moving parts. There is very low power consumption requirement for nearly all types of tuned vibratory feeders and as mentioned before, maintenance is low, particularly for magnetic feeders; even for the rotary vibrator feeders, the only maintenance required is occasional lubrication of the bearings in the motor. The initial cost of the tuned vibratory feeder is quite low and the installation cost is low because there are no special requirements for this feeder, just some suspension cables.

One disadvantage is, that the feed extraction is not fully positive. This means that if extreme accuracies are required such as one half or one percent accuracy, then a weight system of some type may be recommended.

Another problem may occur in handling fine or sticky material; however, this for the most part is overcome by some of the newer designs which offer higher accelerations, higher strokes and ensure that materials will be projected from the trough. The problem that may occur with some very dry fine materials is a dusting condition at the outlet of the feeder. This can be overcome, however, by proper hopper design or if the material is extremely dusty the trough can be covered making the feed area dust tight.

6.1.7.2. Supports for Trough Feeders

The support or suspension of a vibratory feeder must be capable, on one hand, of carrying the entire load, and on the other, of preventing the vibrations being transmitted to the surrounding structure. While the first task requires a firm support, the second demands a flexible, elastic mounting which transmits as little as possible of the dynamic forces to the supports, the bunker, or the building. The natural frequency of these structural elements must not coincide with the operating frequency of the vibratory feeder.

The various types of support or suspension which can be employed are shown in Fig. 124 (F 1):

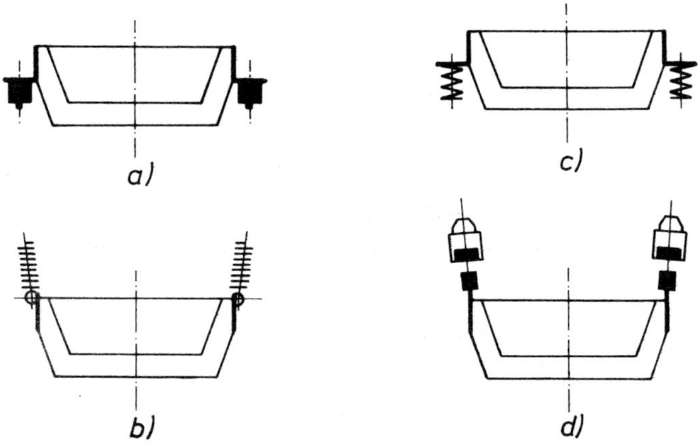

Fig. 124

a) Mounting on rubber blocks.

Advantages: simplest method of support;

very good positioning of feeder, especially during starting up and closing-down; small deflections; great ability to absorb overload.

b) Suspension by helical springs.

Advantages: flexibility, good vibration isolation.

Note: as the helical springs do not hold the feeder in an exact position, it is necessary to make sure that it cannot come into contact with nearby, static structures by moving horizontally.

c) Mounting on compression springs.

Advantages: flexible, good vibration isolation; to be employed in the case of bulk material at high temperatures.

d) Suspension by rubber blocks.

6.1.7.3. Feeder Flow Patterns

There is a natural tendency for vibratory feeders to withdraw from the front portion of a hopper opening, see Fig. 125. To help alleviate this problem,

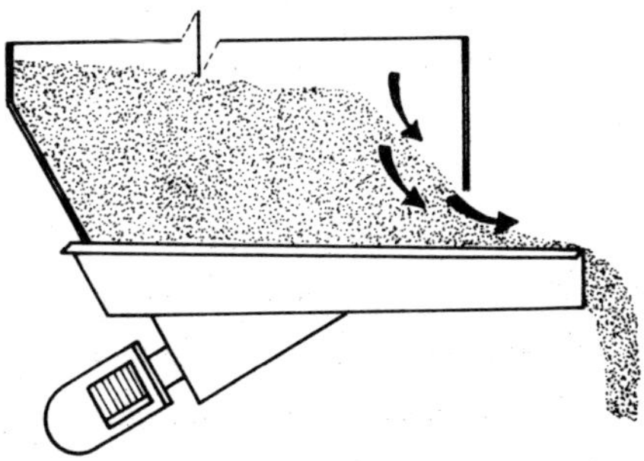

Fig. 125

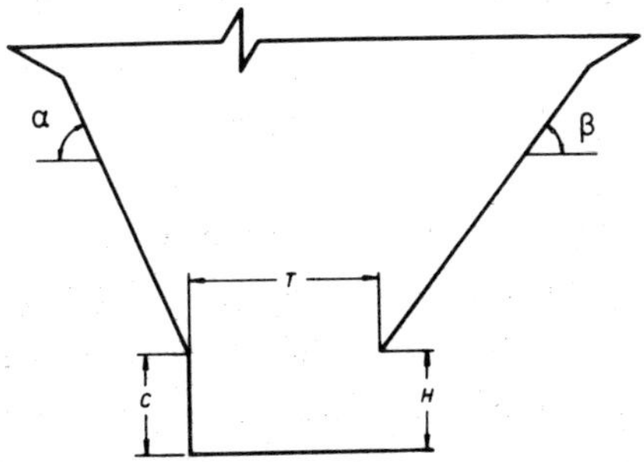

Fig. 126

proper hopper design will permit material to be placed on the feeder at the far end of the pan. Fig. 126 shows a design that does this by supplying a short vertical section just above the rear of the pan, Section C. This length must be at least the height of the pan. A preferred height of one foot is recommended for high tonnage applications. In addition to this the slope angle α of the rear hopper wall should be quite steep permitting flow along the wall. A slope that is just shy of creating flow along the wall will minimize abrasive wear on the wall but is more likely to arch and rathole. The slope of the front wall of the hopper should be just marginal for flow along the walls, 50 to 60° is acceptable for most materials. This permits a large reservoir of slow moving material to be just above the gate - the portion of the hopper that will naturally withdraw the fastest. As a result of the steep rear wall and the shallow front wall the hopper will empty quite uniformly. It is best to have the side walls vertical or at least steep enough to allow flow along the walls.

The dimension T, for randomly sized material, of Fig. 126 should be a very minimum of 2 1/2 diameters of the largest particle size to eliminate large particle interlocking. In applications which have near sized material T should be five times the particle size. To assure there will be no bridging or arching due to cohesiveness of fine particles, the opening should also be sized greater than the minimum arching dimension. This dimension may be determined by referring to references (133) and (137). The diagonal of this opening should be large enough to overcome ratholing - these two quantities are considered in a good hopper design.

The gate height H should be at least twice the size of the largest particle size and adjustable by means of a slide gate. The gate height will increase proportionally to the capacity required and the particle size. An approximate gate height may be determined by the depth of flow at the end of the trough required to deliver the required capacity. The gate height H should be approximately 1.5 d where d is determined by:

$$d = \frac{\text{capacity (ft}^3\text{/min)}}{\text{width} \times \text{flow rate (ft/min)}}$$

Feeder widths are generally chosen to be about 1.5 times the gate height H, depending on capacity requirements. Best flow patterns result if the gate height H is roughly twice the throat dimension T.

Flow velocities vary from 20 to 60 feet per minute depending on the material properties, feeder amplitude and operating frequency. On any given feeder poor hopper design will decrease conveying speed and depth of the flowing material both of which decreases the amount of material delivered.

For relatively high capacity feeders, skirt boards will be required on the sides of the gate opening, extending to the outlet of the trough, see Fig. 127. In order that the skirts retain the material without interrupting or restricting

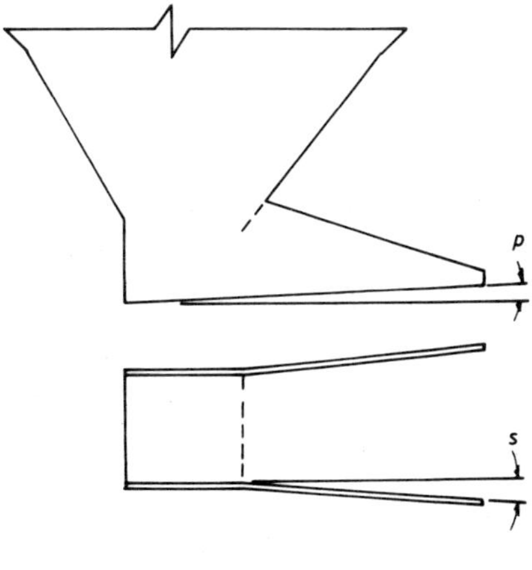

Fig. 127

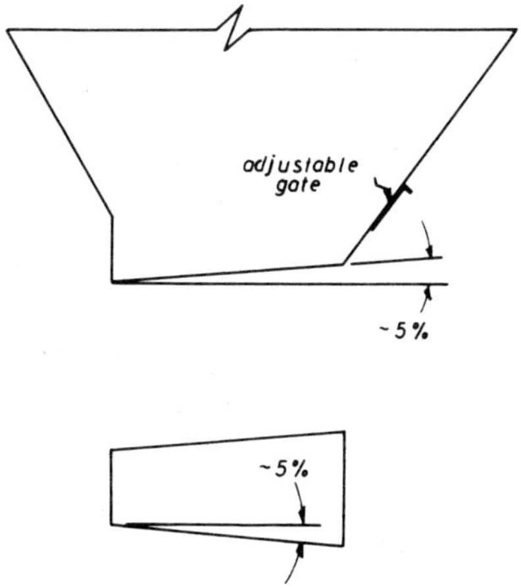

Fig. 128

the flow, they should be tapered away from the bottom of the feeder trough along the length of the trough at a rate of at least one inch per foot of feeder length. This is indicated by the angle p in Fig. 127. In addition the skirts should be wider at the outlet of the pan than at the hopper; this taper should be at a rate of at least one half inch per foot of feeder length, the angle s of Fig. 127 depicts this taper.

When high capacities are required of a feeder the pan is sloped down from 6 to 15°. Some installations have boosted delivery rates by more than 50 % with a 10° downslope. A rule of thumb value is a 3 % delivery increase per degree of downslope. This is obviously the approach to get the most tonnage per dollar investment. By downsloping, the feeder is getting some "free" assistance from gravity. With increasing downslope, however, there is an inherent loss of control. The converse holds true, upsloping the feeder gives better control but lowers delivery rates considerably. Skirts are nearly always required in installations where the feeder is given some downslope. The same recommendations of tapering the skirts hold true as for the level feeders discussed above.

A particular problem is created with materials that will form stable ratholes over a relatively large diameter. These materials often require a long slotted opening to overcome ratholing problems. In such cases care must be taken to design a transition hopper to the feeder that will give fairly uniform withdrawal across the length of the slot. Vibratory feeders are well suited for feeding across the length of such an opening with virtually no limit on size. However, many design conditions require that feeding is done from the narrow end of the opening, resulting in a long narrow trough. To overcome the natural tendency of vibratory feeders to feed only from the front, the transition hopper of Fig. 128 is suggested.

In the horizontal and vertical plane the hopper walls should be diverging at about a rate of 5 % of the length, or about one half inch per foot per side (see Fig. 128). This taper relieves the pressure in the flow direction, allowing material to be placing on the trough at the heel of the trough and increasing capacity along the length. To eliminate flushing on this type of installation, skirts may be added to the sides of the hopper, external to any flow area, see Fig. 129. When skirts are required, their separation should be at least 10 % greater than the width of the hopper gate opening. Tests have shown that a combination of horizontal and vertical taper is most effective. The rear wall in this hopper configuration should be nearly vertical or at least have a short vertical section (at least one foot in height). The front wall may be sloped as shallow as 50° - 60° and should be equipped with a slide gate. With a hopper designed in this manner flow rate will increase linearly proportional to the gate opening, and flow patterns within the bin will improve with larger gate openings. For gate opening of approximately two feet relatively uniform flow patterns can be obtained in the bin, provided other conditions required for mass flow are present i. e., the

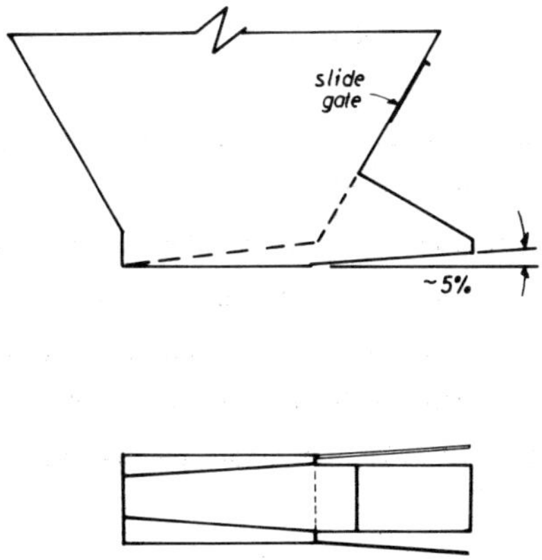

Fig. 129

proper wall angles on the hopper and a head of material above the hopper section.

From the standpoint of feeder operation, the feeding from the narrow end of the long slotted opening has an interesting side effect. There will be some amount of vibration transmitted to the material in the bin. This vibration has two effects: it makes material "live" that would otherwise be stagnant, thereby promoting flow reducing the likelihood of a flow stoppage due to bridging or ratholing; while the material is being made live this requires energy. This energy is not lost in moving the increased head load, but more so from the material damping present due to inter-particle friction and compression of the material. Whatever amount of energy is absorbed by this damping must be put into the system by a large drive. Consequently long slotted openings frequently require more horsepower for the same delivery rate than would be required by a comparatively short but wide feeder. The most efficient type of vibratory feeder installation is one using a nearly square trough. Here, about 25 % of the back part of the feeder is under the hopper outlet, the remaining part of the feeder can be used for conveying and providing for the static angle of repose of the material.

Along with the minimum requirements for throat opening and gate height, the angle of repose of the material, its particle size, and the feeder down slope prescribe the dimensions of the feeder pan and consequently the size of the drive unit for normal installations. Fig. 130 shows a down sloped vibratory feeder with a minimal length trough, when the feeder is not operating the material will flow to Point C because of its natural angle of repose (angle β of Fig. 130). On operation the material has a flow boundary

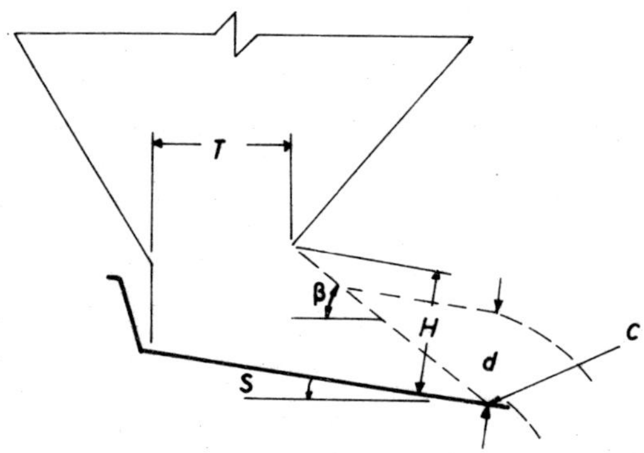

Fig. 130

indicated by the dashed line, the minimum depth d as pointed out before determines the gate height.

6.1.8. Hopper Gates

The simplest type of feeders are gates which are operated by hand levers or, under control of an operator, by air, hydraulic piston or motor drive. They can only be successfully used for the feeding of free flowing materials and in plants where automatic control of bin discharge is of less importance. Therefore, it is not intended to describe in detail the range of operation of each type of gate. Fig. 131 shows schematically the various arrangements of which many variations exist (24), (34), (219), (255), (279):

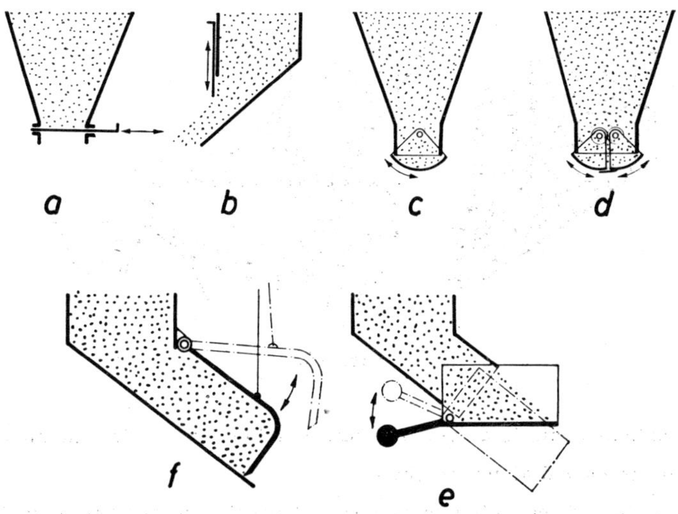

Fig. 131

a) horizontal gate,

b) vertical gate,

c) single quadrant gate,

d) duplex quadrant gate,

e) hinged chute gate,

f) swinging cut-off gate.

Types a, b, c, and d are mostly used for the storage of fine and medium materials, such as grain or gravel, while types e and f find their application for coarser materials like ore and coal.

A special type of gate consists of heavy chains which hang like a curtain on the outlet. This method is applicable only for coarse material (Fig. 132) (219), (255):

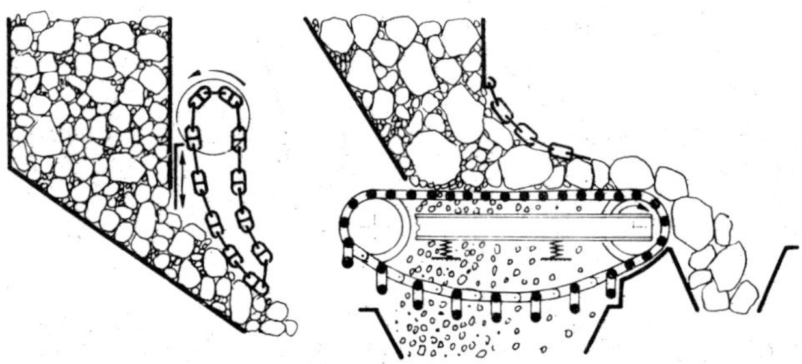

Fig. 132

left: an endless chain is driven by a drum and by this motion the bulk material is moved along the chute,

right: a chain is fixed to the hopper wall above the outlet. The material is discharged by a travelling grate providing simultaneously screening.

6.1.9. Conclusion

The type and size of feeder for a given application is primarily dictated by the customer's specifications of capacity requirement and the characteristics of the material that will be handled. It is imperative that the user states clearly and completely the various characteristics of the material to be handled, and that he appreciates that an overstatement may cause an excessively large feeder to be chosen; an understatement may result in failure to achieve the required capacities by a significant amount.

To assist the reader in making a first selection among the available feeder types, Table 13 compares the various feeder types in relation to the material sizes which are best handled on these feeders. The letter E indicates the feeder is an excellent choice for the application. The letter G indicates that the feeder is good or acceptable and F means that the feeder is fair and that special considerations may have to be taken to handle the type of material indicated on that feeder. The chart indicates that the apron feeder is very well suited to very coarse and coarse materials. With the belt feeder the fine and very fine materials can be handled easily. Rotary plow feeders are most compatible with the medium and fine material. For the rotary table feeder, it appears that only the very fine material would be ideally matched, fines and concentrates can be handled but not as easily. Screw feeders are primarily applied to the concentrates and powders. The vibratory brute force feeders are extremely good where very coarse and coarse materials are handled. This is primarily because of its impact resistance and abrasion resistance. With fine materials the brute force feeder has a disadvantage because it does not offer a refined control system. The vibratory tuned feeders are well suited to the very coarse materials down through the fine materials. With concentrates special considerations may have to be taken. This is particularly true if the concentrates are extremely wet. With very fine powders in high tonnage installations it must be remembered that the flow depth is very shallow (five to eight in. at most).

As it is difficult to answer the question "what feeder should be used for a bulk material?" without any restrictions, the Table 14 has only the purpose

Table 13

feeder type	maximum material size: very coarse + 12 in.	coarse - 8 in.	medium - 2 in.	fine $-\frac{1}{2}$ in.	very fine - 100 mesh	concentrate - 200 mesh
apron	E	E	G	F		
belt		F	G	E	E	G
rotary plow		F	E	E		
rotary table				G	E	G
screw			F	G	E	E
vibrating (brute force)	E	E	G	F		
vibrating (tuned)	E	E	E	E	F	F

of showing the types of feeders that are commonly used (24), (171), (311). The various bulk materials are characterized by a "factor of difficulty" in the last column, which is related to flowability. The number 1 indicates easy flowing and number 10 very difficult flowing materials. This simplified classification is based on research work combined with practical experience. The check mark means a satisfactory application.

Another comparison is shown in Fig. 133 indicating the horsepower required per hundred tons delivered, based on the capacity in tons per hour. This value decreases with increasing capacities. According to the figure the rotary table feeder requires the greatest horsepower per ton delivered. The screw feeder is also quite high but these are speciality applications so this is not really of great concern to us. The high tonnage feeders, the apron feeder, the belt feeder, the rotary plow feeder, vibratory brute force and tuned feeders may be of more interest in comparison. The rotary plow

Table 14

material	apron feeder	belt feeder	scraper feeder	screw feeder	vibratory feeder	rotary vane feeder	chain gate feeder	push feeder	rotary table feeder	rotary plow feeder	factor of difficulty
aluminum oxide			x	x		x					4 - 5
amonium sulfate		x			x					x	6 - 8
ashes (dry)		x	x	x	x	x			x		2 - 3
ashes (moist)		x									5 - 7
barite	x	x	x		x						2 - 5
blast furnace slag		x			x				x		2 - 3
cement			x	x		x					6
cement clinker	x	x			x				x		2 - 3
chips			x	x							9 - 10
clay	x	x			x				x	x	7 - 8
coal (normal)	x	x	x	x	x			x	x	x	3
coal slime	x	x		x	x						5 - 6
coal dust			x			x					6
coke	x	x			x					x	4 - 5
dolomite	x	x			x			x	x	x	2 - 5
fireclay	x		x	x		x			x		2 - 5
flue dust			x	x		x					7 - 8
grain				x	x						1 - 2
gravel		x			x				x		2 - 3
iron sulfate		x			x						6
limestone	x	x			x		x	x	x	x	2 - 4
magnesite	x	x			x			x	x	x	3
ore (heavy)	x				x		x	x		x	4
ore (light)	x		x		x			x		x	4 - 5
sand (dry)		x			x			x	x	x	2
sawdust		x	x	x							5 - 7
soot			x	x		x					7
titanium dioxide				x		x					6
zinc oxide				x		x					6 - 8

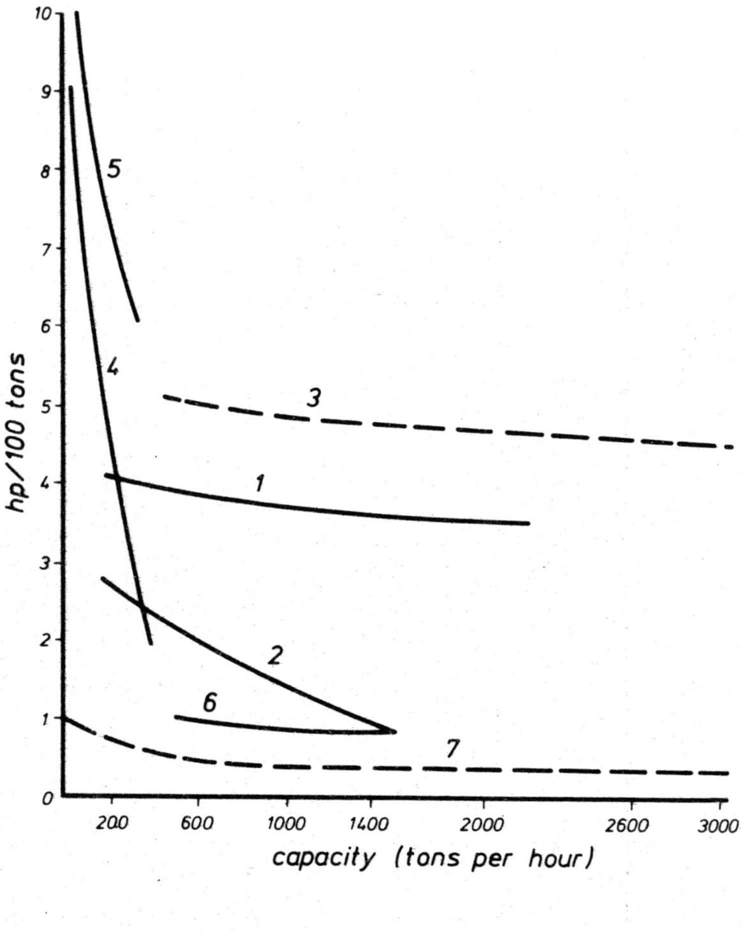

Fig. 133

1 = apron feeder

2 = belt feeder

3 = rotary plow feeder

4 = rotary table feeder

5 = screw feeder

6 = vibratory feeder, brute force

7 = vibratory feeder, tuned

feeder apparently has the highest horsepower requirement - it is about five horsepower per 100 tons delivered. This is because of the high torque requirements to rotate the plow on the shelf. For the apron feeders the horsepower requirements are low, between two and four horsepower per hundred tons delivered. These horsepowers are higher than the belt feeders and the vibratory feeders because of the weight of the apron feeder pans and the friction at the many bearing surfaces.

Now the low horsepower per ton delivered feeders are considered. These are the belt feeders and the two vibratory feeders - the brute force and the tuned system. The belt feeder requirements are between one and three horsepower per hundred tons delivered. The brute force vibratory feeder require about one horsepower per hundred tons delivered while the tuned system requires only between one half and one horsepower per hundred tons delivered which, of course, diminishes as tonnage requirement increases. These are rule of thumb values and for a good properly sized installation they are usable values.

Fig. 134 shows an approximate comparison in dollars per tons delivered based on the capacity of the feeder. There are some definite inflection points and cross over points here. One thing that must be understood at the outset is that the values here are based on a good installation using a 100 pound per cubic foot material. For the belt feeders and the vibratory feeders, there is a definite cross over point which may be of interest in the high tonnage areas and it shows that the costs for the belt feeders are still diminishing while the vibratory feeders roughly reach a constant value. Belt feeders are capable of handling up to 2, 000 tons per hour and this is not extended further than that on the chart. For the apron feeders and rotary plow feeders, the costs are still not out of hand and, for given installations, certainly all of these feeders should be considered. Rotary table feeders and screw feeders are high cost feeders, but are speciality

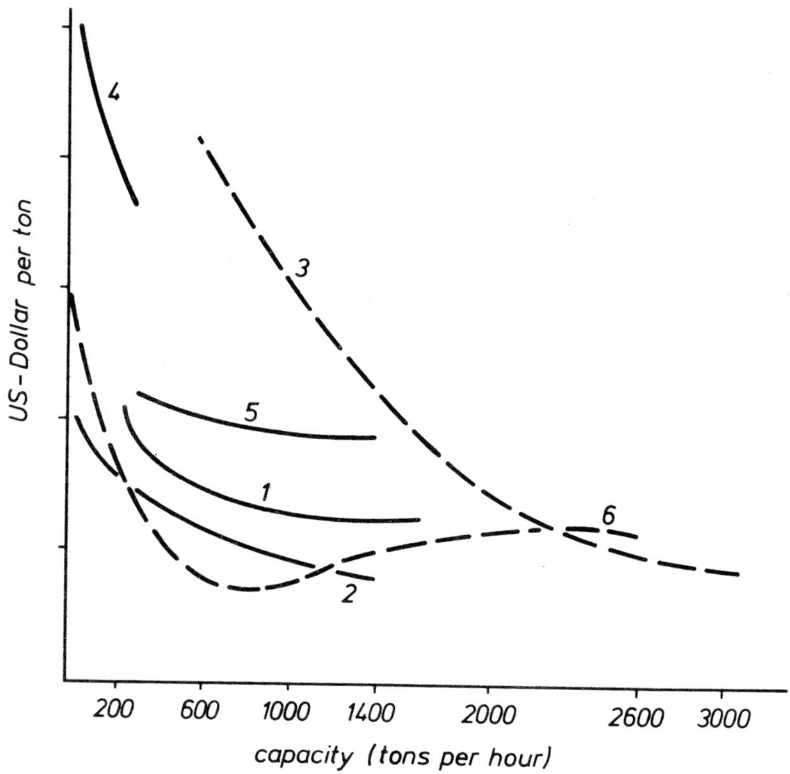

Fig. 134

1 = apron feeder
2 = belt feeder
3 = rotary plow feeder
4 = screw feeder
5 = vibratory feeder, brute force
6 = vibratory feeder, tuned

application feeders. The costs shown are representative values and should not be the primary factor in selecting a feeder.

In summary, then, all feeders have advantages and disadvantages; cost comparison, horsepower comparison have been given. It has been pointed out which material sizes are compatible with the various types of feeders; the capacities, advantages, disadvantages and limitations of the feeders have been pointed out. For a particular installation these feeders are to be evaluated by considering cost, maintenance requirements, space requirements of the feeder, the control requirements for the process, the accuracy of flow regulation, the uniformity of flow and many other factors.

6. 2. Aeration Procedures

After having defined the flow function, J e n i k e (133) could classify (see Chapters 1. 3. and 2. 1. 4.) the fine grained bulk material in a simple way. He established limiting values of the flow function $FF = \sigma_1 / f_c$ as the parameter of classification. The bulk materials with FF values of 2 and less are non-free flowing or sluggish (268). There are a number of fine grained materials, which tend to fluidize, for which C a r r (64), (65), (66) has established criteria of classification. Such materials can only be handled in special bunker constructions. The difficulties in discharge of the non-flowing materials are caused by large cohesive forces between the individual particles, which increase with the consolidation of the material. In the case of fluidizing materials the problems arise from the expansion of the air trapped in pores as well as the ability of the air to be adsorbed at the particle surface. Because the bulk density can fluctuate over a large range, further difficulties arise during volumetric or gravimetric feeding (163), (201).

Such problems are mostly solved with the help of aeration techniques, which can principally be applied to all powders. By blowing air in the bottom region of the bunker, the materials will expand, so that the close contact between the grains is loosened and they can be transported with a constant flow rate. As the entire bunker cross-section can be activated, according to the first-in/first-out principle, the surface of the bulk material can be lowered uniformly and thus a mass flow can be achieved. As the movement of the bulk material towards the outlet is no longer regulated by the gravitational force alone, but primarily by gaseous media, usually air, storage facilities can be constructed as flat bottomed silos with diameters of 30 meters and more (F 13).

The flow rate is primarily influenced by the blowing pressure. Engh (90) has proved, that a Bernoulli type equation of energy can be accepted as a formula to describe a material flow. This function can be approximated as:

$$G/A = const. \cdot \sqrt{\Delta p}$$

where G = flow rate

A = orifice area

and Δp = gage pressure.

Similar to Carr, Schmidt-Holthausen (264) and Bruff (61) divide the fluidibility of powdered materials according to the angle of repose of the fluidized solid. It is characteristic for the material properties such as grain size, grain distribution, grain form, surface properties etc.

Fig. 135 (264) illustrates the fluidibility of two materials as a function of the amount of air. As a further criterion it can be mentioned, that the better a material can adsorb air and retain it on the surface or in the pore space, the better it can be pneumaticly transported. In Chapter 3 it was noted, that the fluidized materials achieve the hydrostatic pressure conditions. The difference of the wall pressures after shutting off the aeration i. e. their behavior during the ventilation can be taken as a standard of comparison (Fig. 136) (264).

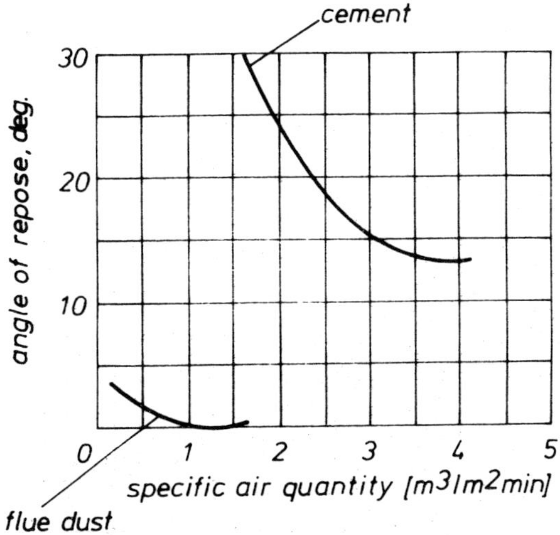

Fig. 135

For the form of the bunker and for the arrangement of the aeration elements various suggestions have been made. The entire bottom can be aerated, when porous materials are installed between the bulk material and the pressure chamber, such as porous ceramic tiles, woven cotton or terylene belting, sintered metal (88), (108). The material must be uniformly porous, but at the same time dense enough to cause considerable restriction to the air flow, so that in the case of heterogeneous capping of the bottom the air does not escape at the points of minimum material resistance. The bottom form can be designed as the usual bunkers (88). For smaller plants of conventional types, it is sufficient to blow the air only at some points into the hopper.

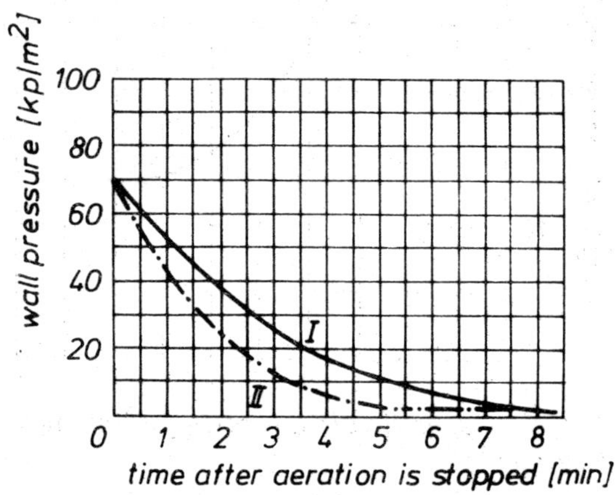

Fig. 136

Especially in Germany a number of bunker types for pneumatic discharge were developed for large installations with diameters of 30 meter and heights of 30 to 40 meters (96), (201), (264), (F 13), (F 43), (F 48). The aeration is realized with chambers or chutes in the bottom, making it possible to control the air stream and the air pressure.

Figures 137 (264) and 139 (237) show some developments, starting from conical saddles between the aeration chutes. In the case of such saddle constructions (Fig. 137, I) it is possible, that a flow channel is formed between the saddles and that the material bridges over them.

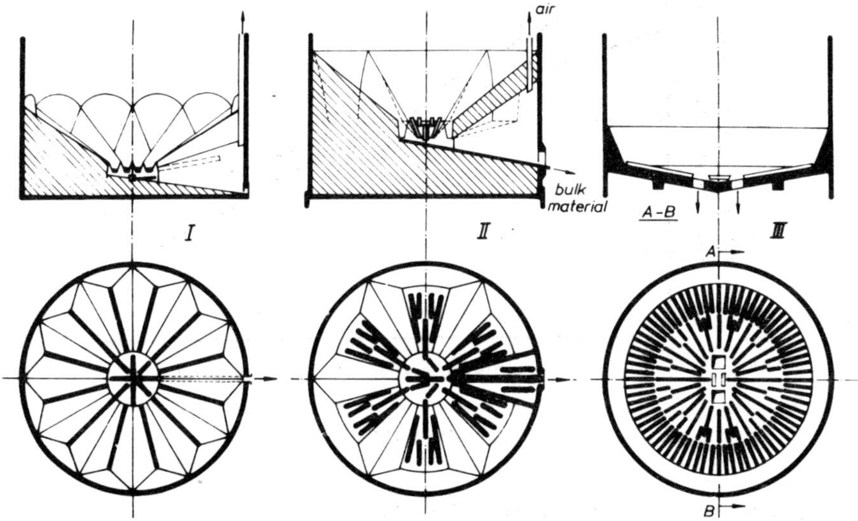

Fig. 137

It is customary to blow uniformly a specific amount of air per min through the bottom area (264). In the case of radially installed elongated chambers, the ratio of the aerated and non-aerated area decreases greatly with the distance from the silo outlet (Fig. 138, line I) (264). In order to achieve a constant rate of air flow over the whole cross-section, a complicated regulation and control of air pressure is necessary.

Because of this reason, smaller aeration units which can be distributed uniformly over the entire bottom cross-section are preferred nowadays (Fig. 137, III and Fig. 138, III). The specific air consumption lies in the

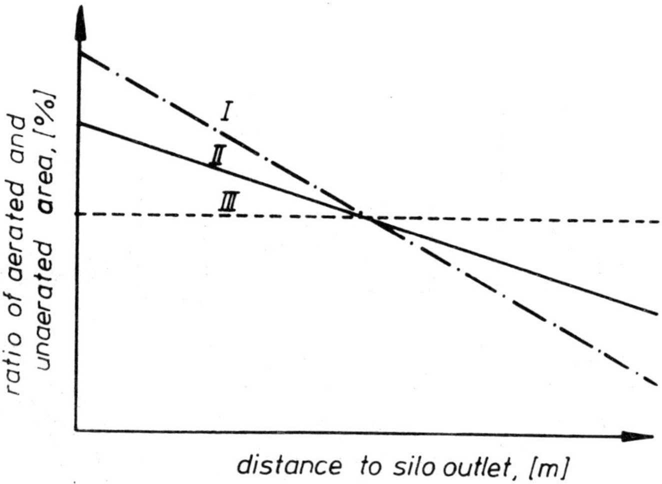

Fig. 138

magnitude of 3 m^3 per min. per m^2 of the bottom area.

As it is difficult to transport the fluidized materials in the discharge conveyors, an expansion chamber is very often installed in the silo, where the material can be ventilated before being discharged. In the chapter on homogenization, the Mixing Chamber Silo was already mentioned (Fig. 139). The expansion chamber is installed concentrically over the aeration area. During discharge the aerated material within this chamber is not compacted by the overlying bulk material and thus an arbitrary loosening of the material can be achieved (237), (F 13). In order to minimize the expenditures for the aeration as well as to reduce the number of aeration units in the case of larger bunker diameters, a cone is very often attached to the bin wall.

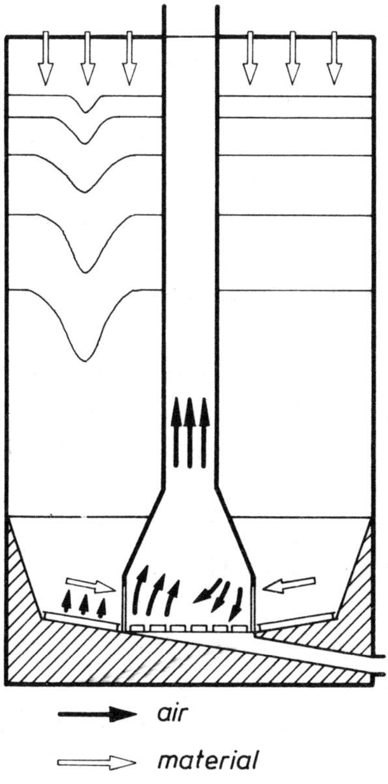

Fig. 139

The inclination of the bottom towards the discharge must be suited to the fluidibility of the stored material. No generally valid information about this is available.

7. Functions of Bins and Bunkers

As has been discussed in detail in previous chapters, bins and bunkers are normally used for the storage of bulk material. However, there are also some special functions they may have to fulfil. In many chemical processes it is necessary that a granular solid substance has to react with a gas for a certain time. In this case bins are often used as reactors in which the bulk solids move downwards in mass flow at a constant rate and a counter-current gas flow interacts with the solids.

Another example of bins used as reactors is related to a nuclear process. The core of a pebble bed reactor consists of fuel elements which are balls of equal size. The pebble bed is in a state of a continuous circulation, and mass flow is required with absolute accuracy. Special investigations on the material movement were conducted by (38), (39), (106). Some of the results have already been discussed in Chapter 1.1.

Furthermore, the flow problems in shaft kilns and blast furnaces can be compared to some extent with those of bins. Some of the results of research work on bins have been transferred successfully to such facilities to achieve a better flow of the material (99), (151), (162).

The fundamental laws for the gravity flow of granular material are also applicable to the mining industry. This is important for the block-caving mining method in stable ore deposits, such as iron ore (Fig. 140) (109), (127). In this method the ore of a finite block size is blasted into a cavity within the deposit from which the ore is then discharged. Considering this cavity as a bunker, the problems of storage and withdrawal equal to those in real bunkers and bins. The purpose of this mining method is that the blasted ore is stored and that from the level of the broken ore holes are drilled for the next blast. The discharge rate can be as high as required.

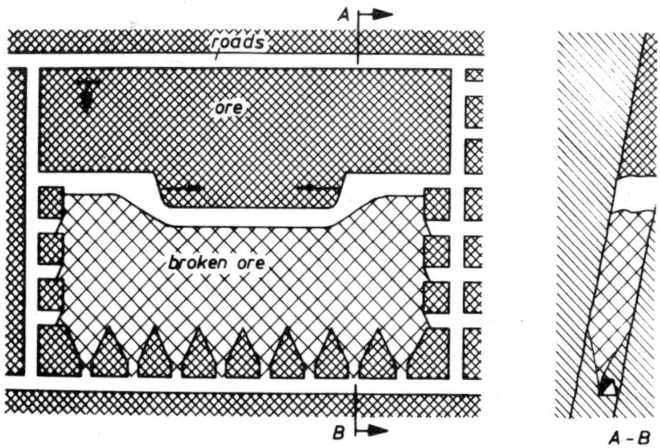

Fig. 140

The number of drawpoints is important for the degree of segregation.

Besides these general descriptions of functions of bins and bunkers, three others will be considered more intensively in the following paragraphs, because of their importance in technical processes.

7.1. Homogenization of Stored Material

In order to achieve constant input conditions for production processes, it is very often necessary to compensate for the deviations in the quality of the raw material stream or to blend different materials with each other in order to retain a mixture within a narrow range. In many cases it is possible to achieve sufficient homogenization in a bunker providing the equipment for charging and discharging is specifically designed for this purpose. These types of blending bins, as far as these have become known through publications, will be discussed in this chapter.

Similar to the evaluation of segregation (Chapter 8.1.2.), the standard deviation σ can generally be utilized for the evaluation of blending.

$$\sigma = \sqrt{\frac{1}{n-1} \sum_{i=1}^{n} (x_i - \bar{x})^2}$$

n = number of samples

X_i = measured values i. e. % of one component

$\overline{X}$ = desired value

The efficiency of blending can be approximated from the ratio of the standard deviation before and after the blending processes σ_1 / σ_0 (Fig. 141). The blending efficiency can be related to time by means of an e-function, in the form of $\sigma_1 / \sigma_0 = e^{-A \cdot t}$

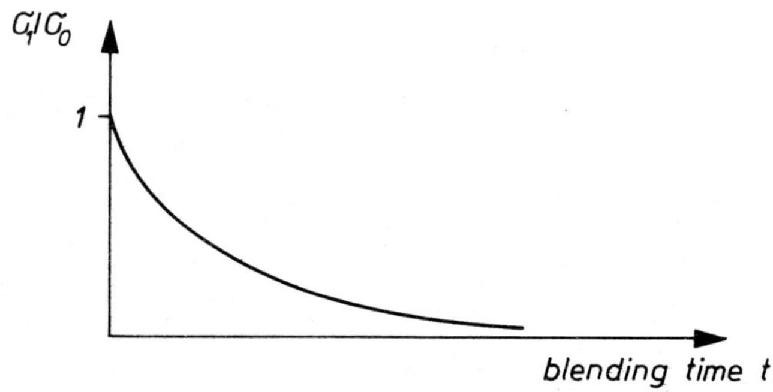

Fig. 141

The constant A is a measure of the intensity of blending and t is the duration of blending (319). The above function ranges in values between 1 and 0:

$$t = 0 \qquad \sigma_1/\sigma_0 = 1$$

$$t = \infty \qquad \sigma_1/\sigma_0 = 0$$

The value σ_1/σ_0 = o for t = ∞ is a hypothetical one, not only because blending methods with larger durations are uneconomical, but because a perfect blending i. e. an ideal and uniform spacial distribution cannot be achieved. Using the probability theory, Pawlowsky (222) has shown that unavoidable stochastic deviations occur.

Weislehner (312), (314) considers the constant of blending A not to be a significant standard of comparison for different blending methods. Therefore he introduces the "half-value time", commonly used in physics, in the above equation. T represents the period of time necessary to reduce the standard deviation σ_0 to half its value.

$$\sigma_1/\sigma_0 = 1/2 = e^{-A \cdot T}$$

$$T = \ln 2/A$$

In order to compare not only the blending efficiency within one bin, but also the effective efficiency between different blending procedures, Weislehner has further defined the specific blending energy N_s as

$$N_s = N \cdot T/Q$$

where N = power consumption

Q = weight of the mixed material

In the case of determining the time necessary to achieve the required blending efficiency, the superimposing effect of segregation must also be taken into account. In such cases the maximum value for blending efficiency at t = ∞ is no longer σ_1/σ_0 = o, but any other value < 1, which can be determined by the intensity of segregation. Above this optimum no improvement of the blending can be achieved, because the blending and segregation compensate each other. This compensating effect has been recognized by Weydanz (316) utilizing a mixing drum. The time function for the blending efficiency and the optimum blending for a system must be changed accordingly to:

$$\sigma_1/\sigma_0 = \sqrt{e^{-2At} + (\sigma_\infty/\sigma_0)^2 (1-e^{-2At})^2}$$

In the above expression the first term under the square root defines the mixing process and the second term the segregation tendency.

Pneumatic homogenization has become one of the most important methods for mixing fine grained and powdered material. It represents a special method and will be discussed only briefly, because such installation mainly serve for blending and not for storage. The principle consists of feeding the bin in batches and of circulating the material in the bin by blowing air through porous sections of the bottom for a certain blending time t. During this process only certain bottom sections are aerated and not the whole bin bottom. Wiegmann (319) has compared four methods commonly applied in the industry (Fig. 142).

The shaded areas are aerated and after a certain time this is changed over to the other sections. In these blending bins batches up to 2,500 tons can be mixed in one operation. For details on the theory of pneumatic mixing and for further information about experiences in operating plants, referenced literature can be consulted (54), (89), (146), (153), (174), (187), (189), (226), (236), (258), (313).

Analogous to the classifying and segregation effect of a bunker, one can also talk of a homogenization effect. During a core flow zone is formed above the outlet, all horizontal layers within the core are very quickly activated. The flow patterns, illustrated in Fig. 1 (Chapter 1.1.), show the core flow effect in a bin filled with layered materials, and the effect on blending after a crater is formed at the top surface can be recognized clearly.

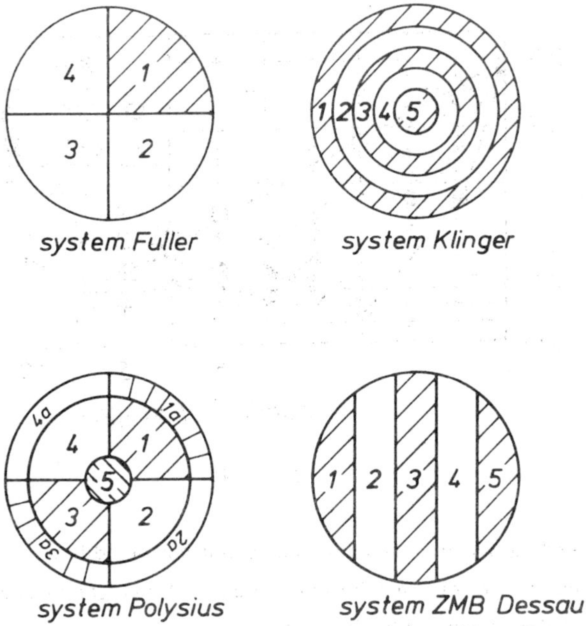

Fig. 142

It is evident, that the blending effect is influenced greatly by the geometry of the bunker. The definition of mass flow states, that the surface moves down uniformly. Accordingly only a smaller homogenization effect is to be expected. Generally, the blending effect will increase with decreasing bunker activity (ratio of flow channel volume to the whole volume). The experimental results of Tanaka and Kavai (289) are given in Fig. 143, which show the standard deviation of a bulk material fed in two layers of different granulation. σ is plotted against the number of stages for different hopper slopes. These results may lead to the following generalization:

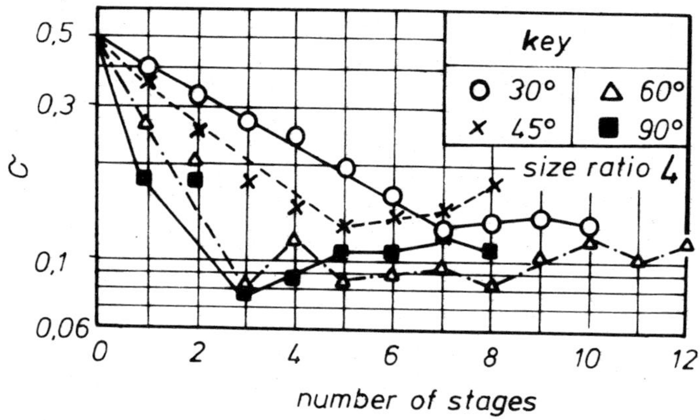

Fig. 143

a) The blending optimum is achieved by flat bottomed bins.

b) For such type of bins a minimum number of charges is necessary to achieve the optimum blending effect.

c) The description of blending processes by an e-function seems to be justified also for such blending techniques.

d) The superimposing segregation effect is evident. σ increases after the blending optimum is reached.

A remarkable improvement of such a multistage blending method (for the time being still in an experimental phase (289)) could be achieved, when suitable measures are taken to reduce the superimposed segregation tendency (see Chapter 8.1.3.).

Similar tests were run by Johanson (254) circulating the bin volume continuously. In this case mass flow conditions must be met because non-flowing regions near the bin wall do not take part in the blending process.

On the other hand there must be a sufficient difference in retention time for solids in different horizontal positions in the bin (see also Chapter 1. 1.).

The utilization of the core flow effect to improve the homogeneity of a bulk material is already realized in the so called "Mixing Chamber Silo", which is equipped with a pneumatic discharge system (Chapter 6. 2.) (237), (F 13). In these silo types, of which a model is illustrated in Fig. 144 (237), the aeration and thus the discharge does not take place uniformly over the entire cross-section, but only in a narrow and limited sector. According to the aeration time of these sectors, the layers of the aerated flow zone move relative to the non-aerated sectors. The material is fed into the bin uniformly over the cross section. In order to get an optimum blending, the time of aeration for the individual sectors must be suited to the specific property fluctuations of the material. As the "Mixing Chamber Silos" are primarily meant for storage, it would be uneconomical to circulate the contents of the silo a number of times to improve the blending efficiency.

To mix different materials or to equalize the quality of the bunker charge the so called "Blending Stack Pile" system can be applied in large plants (144). The slot bunker is divided into two compartments (Fig. 145). The bunker can be charged arbitrarily somewhere between the center and the wall of the bunker. Due to the natural slope of the material an inclined layering occurs in the bunker, which can be varied to quite an extent by changing the charging point during the feeding. For a coarse mixture a rotary plow may be used for withdrawal, which cuts all the layers in rapid succession along the slot (161) (Chapter 6. 1. 4. 2.).

Otherwise a number of individual discharge outlets are to be designed, so that the layers from all regions are simultaneously discharged onto the collecting conveyor.

Fig. 144

The system of "Blending Stack Pile" is modified by the construction of a so called "Organ Pipe Bunker" (144), (282), (F 60) (Fig. 146). The slot bunker is divided into a number of compartments. The individual cells are now fed in horizontal layers.

Because the storage capacity of the compartments decrease stepwise, the blending effect can be achieved through different durations of the individual runs. A plant of such type can be constructed with different variations. The walls of the compartments may run the full height of the bunker or they may begin at a certain elevation in the bunker. In addition, a plant of this type

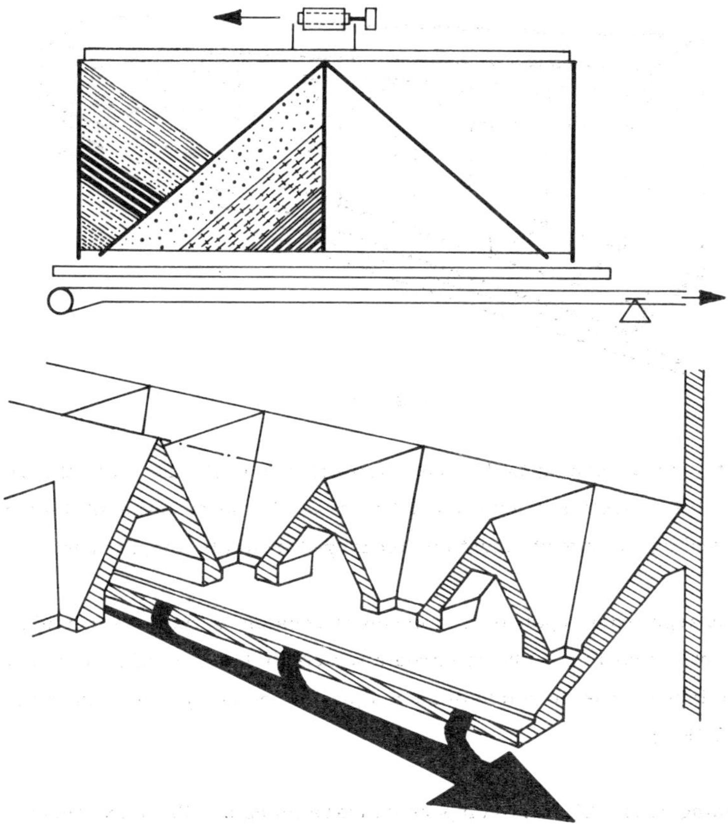

Fig. 145

can also be designed as a slot bunker, when the partition-walls reach only the top of the discharge cones. The most important criterion for such blending bins is the horizontal lowering of the material surface, so that the mixing process can be controlled to some extent (144). The prerequisite for this is the requirement of mass flow, to be achieved by designing the

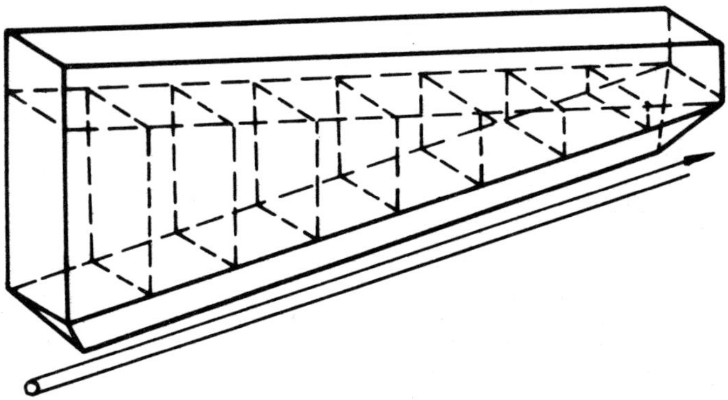

Fig. 146

compartments according to the mass flow criterion. Furthermore the capacity of the collecting conveyor must increase in the direction of the flow stream and equal amounts must be discharged from all compartments.

The discharge from separate bins makes it possible to feed the individual components according to the required mixture, in the case of a multi-component mixture. The principle of such a proportioning system is shown in Fig. 147 (F 1).

The feeding of the different components takes place by vibratory feeders and the material is collected in a common conveyor. With the layering of the material on the conveyor, there is still no real mixture prepared. For an effective mixture, special equipment must be introduced to mix the layered conveyor stream. A mixing device of this type for power station coal, requiring a low capital expenditure, has been described by Aumund/Mechtold (24), and is shown in Fig. 148.

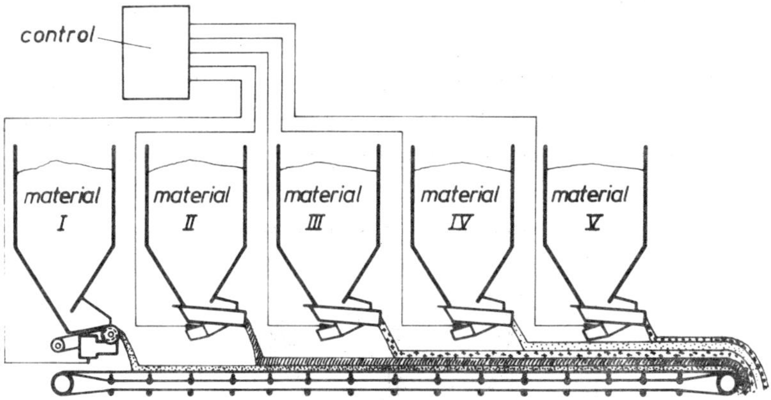

Fig. 147

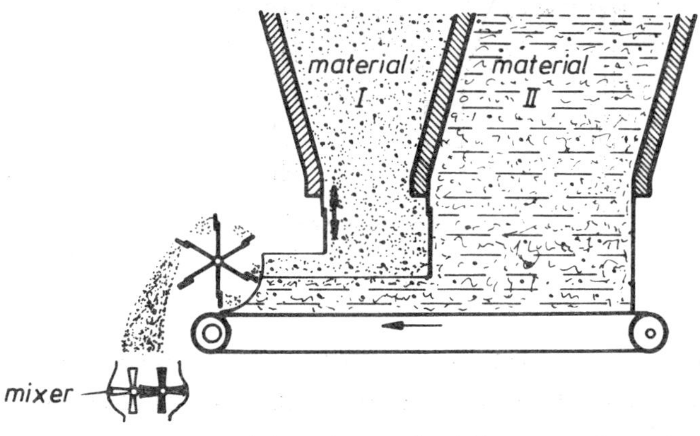

Fig. 148

7.2. Dewatering of Stored Materials

For the cleaning of sand and gravel, which are used for building purposes, the dewatering of the slurries requires considerable installations. The water content of the washed sand may reach up to 20 % which must be reduced before shipment.

One method of dewatering fine sand of grain sizes between 0,1 - 1 mm is illustrated in Fig. 149 (277). The dewatering pipes on the bottom of the bin are covered with gravel layers of different grain sizes to prevent the slot

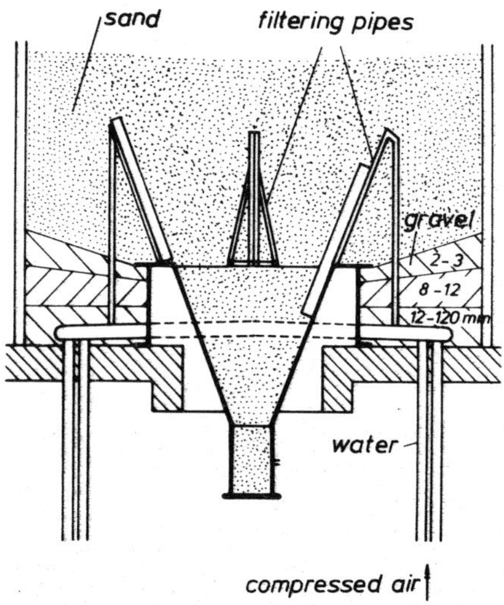

Fig. 149

in them from clogging. They should be connected with a compressed air system for cleaning. Several filtering pipes (well points) are fitted in the hopper which suck the water out of the sand by depression. The same method may be used for sand with grain sizes plus 1 mm but without the filtering pipes.

For dewatering washed gravel the bin may be fitted with a filter cloth on a wall in the region of the discharge opening (Fig. 150) (277). The water

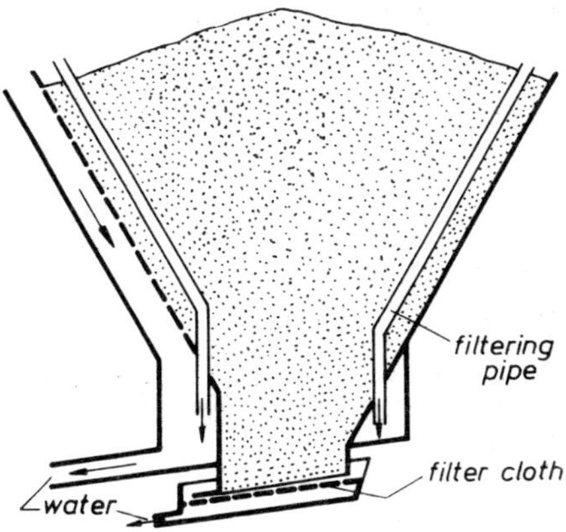

Fig. 150

runs through filtering pipes into a collecting channel.

7.3. Bunker Conveyors

As a simple conveying system is inherently inflexible to cope with material surges, some form of storage is required to even out the material flow. Usually vertical bunkers are suitable for this purpose, but for underground mines where such bunkers are expensive and may cause degradation, another type has been developed, which has gained very wide popularity. Various types of horizontal bunkers or bunker conveyors are illustrated in Fig. 151 (53).

Type A consists of a heavy duty plate conveyor with side plates, providing a cross sectional area for a capacity of 100 to 300 tons of coal. Above it and to one side is a conveyor belt equipped with a traversing plow to deflect the incoming coal into the bunker. The stored material is discharged by running the plate conveyor (F 31), (F 56).

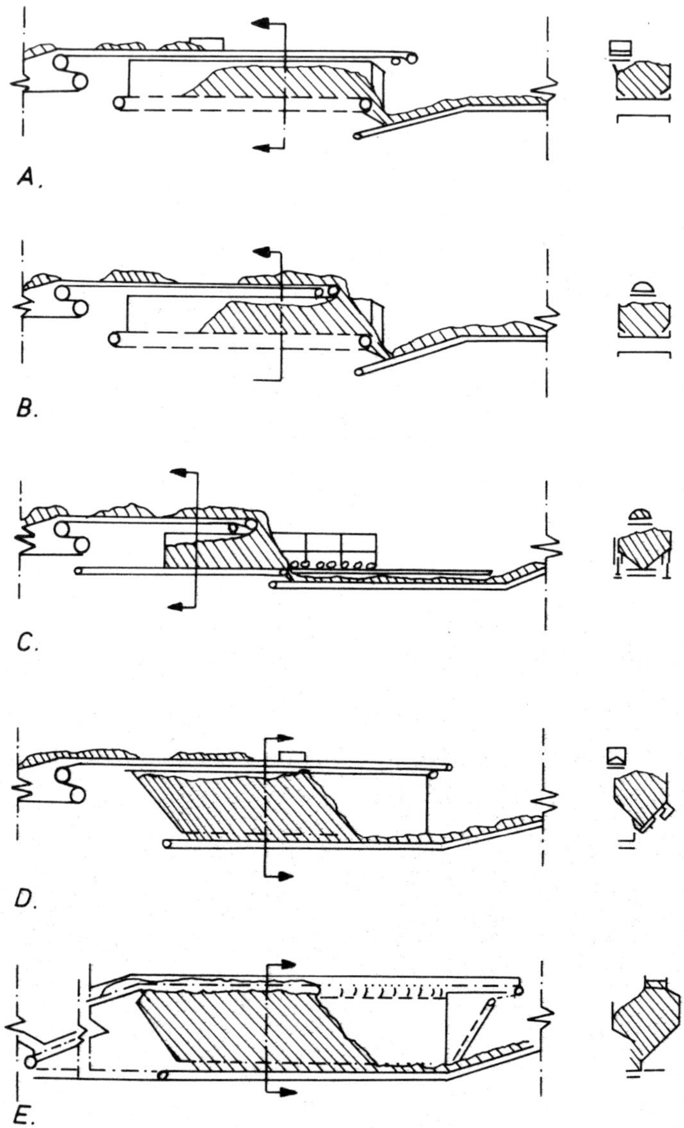

Fig. 151

Type B is similar to A in that the material is moved between fixed side plates by means of a conveyor bed. The head pulley of the delivering conveyor is located near the discharge end of the plate conveyor. When bunkering, the plate conveyor belt is driven backwards to carry material into the confines of the bunker. When feeding out, the motion of the plate conveyor is reversed (F 7), (F 16), (F 31), (76), (310).

Type C is similar in principle, but the high-sided plate conveyor is replaced by a string of open-bottomed cars which can be traversed back and forth. The floor of the train is formed by a flat belt which is pulled along a fixed roller bed as the train traversed (F 17).

Type D is based upon a fundamentally different principle. It consists of a longitudinal trough with a series of hydraulically operated doors along the bottom with discharge onto a conveyor running beneath them. The bunker which can stock as much as 1, 000 tons of coal is filled by a plow, as type A (F 17), (F 56).

Type E is a variation of type D. In this case the bunker is filled by a flight-conveyor arrangement (F 17).

8. Treatment of Bulk Materials

In most of the industrial plants it is important that stored bulk materials do not segregate when they are charged into or discharged from the bin. Degradation of bulk materials during handling and storing may cause additional problems. These phenomena should be eliminated as much as possible, as the change of the particle size and size distribution within the bin may considerably influence the technical process after discharging. Another important factor is the prevention of wear or abrasion of the inside wall of the bins.

8.1. Segregation

Many undesired classifying and separating effects termed as size and density segregation occur during handling of bulk materials in bunkers. They cause changes or variations of material quality important for further processing. Holmes (123) has illustrated both phenomena in a diagram (Fig. 152).

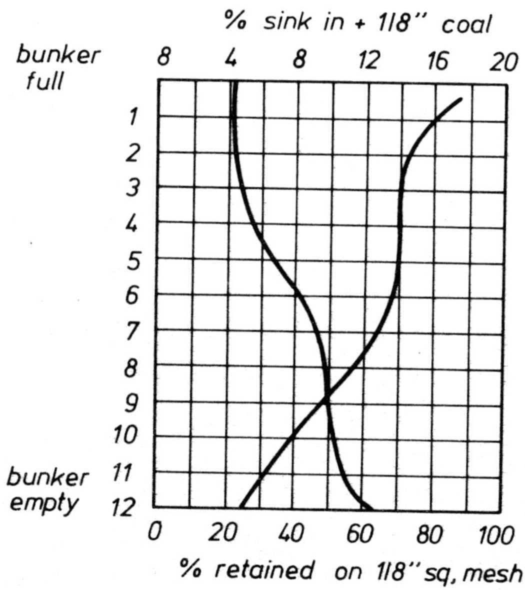

Fig. 152

The percentage of coarse materials in a screen analysis as well as that of heavy materials from a float and sink analysis are plotted against the discharged sample. At the beginning of the discharge process the fine or heavy components are dominant and vice versa.

8.1.1. Cause of Segregation

Segregation is caused by a series of physical processes during charging and discharging of hoppers. Their combined effects can either increase or decrease the total effect. Basically one must differentiate between phenomena at the surface and in the bed of a bulk material. In both cases different mobilities of individual components can exist which are discussed below. For simplification it will be assumed that two fractions of equal density or equal size distribution are present.

During the filling of a bunker a cone is build up, the shape of which is determined by the filling method. Particle movements take place only on the surface of the cone. In case of powders, Williams showed, that only a layer of 2 - 3 particle-diameter thickness is in motion. Following the analogy of a motion on an inclined plane, Matthée (195), (196) has investigated the case of size-segregation. He pursued further the idea of Brown (55) and Tarjahn (290) and established formulas to distinguish different mobilities of smaller and bigger particles. He explains this segregation in terms of differences in size alone and does not introduce different energy levels, because the dimension of mass is cancelled in the equations. As a purely translatory motion cannot cause different accelerations, a rolling motion, at least partly, must also occur. Holmes (123) and Meyer (197) have indicated the sieve effect at the surface which increases in the direction of the bunker wall (Fig. 153). In case of a particle size-ratio of 0.1547 it is theoretically possible that the smaller particles settle in the pore spaces between the bigger ones.

The same authors explain the density segregation as a gravity bedding on the inclined surface. It reduces the mobility of the heavy material and causes a concentration of these components to the center of the bunker.

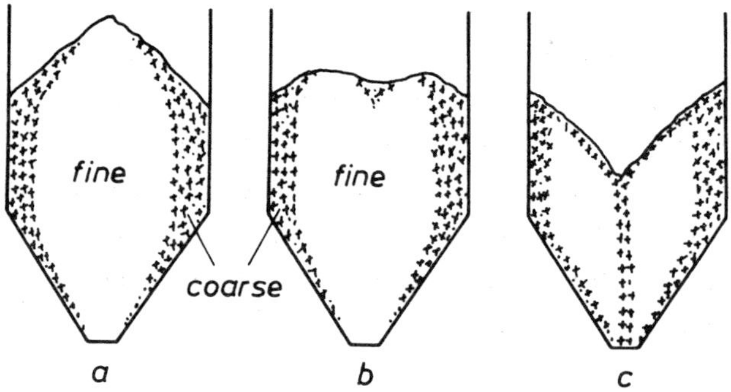

Fig. 153

B r o w n (55) regards the free fall of a particle stream as the main cause for segregation. Further investigations must incorporate, however, the influence of the different elastic properties of the materials. J e n i k e (131) has examined the segregation during free fall in combination with a horizontal component of velocity, assuming the filling of the bin with a conveyor or by an inclined chute. He defines a special material factor W/K (W = weight of the particle, K = coefficient of air-resistance) and claims, that materials with components of widely differing coefficients W/K will tend to segregate under above mentioned conditions. According to J e n i k e , the segregation on an inclined plane can also be explained with the help of the factor W/K, thus contradicting the phenomena of density segregation. T h e i m e r (295) showed that the airflow, created by the material stream in bunkers, is diverted towards the bunker walls so that the light components are carried outwards.

In Chapter 1.1. the flow of bulk material was characterised by mass and core flow. Ideal mass flow i. e. uniform lowering of the material surface, eliminates the segregation effects to a great extent (134), (268). Horizontal layers are discharged successively. Although being segregated in a cross sectional segment, the discharged material has the same overall composition as before.

Core flow, or different radial particle velocities, lead to segregation, because the central core discharges first and contains mainly fine and heavy particles (Fig. 153). The regions near the wall start flowing only after the core has been discharged. These zones are shifted towards the center of the hopper, which analogous to the building up of piles results to a certain remixing.

It remains to be seen to what extent the processes inside a flowing bed lead to different mobilities. Specific results about this are not reported in the literature so far, although, to a certain extent, results of similar studies could be applied. Any motion in a bed of granular material may lead to particle rearrangement and to a layering of the bed. In case of density segregation it tends to achieve a state of minimum energy level. In mineral preparation plants, for example, this effect is used in jigs (Fig. 154).

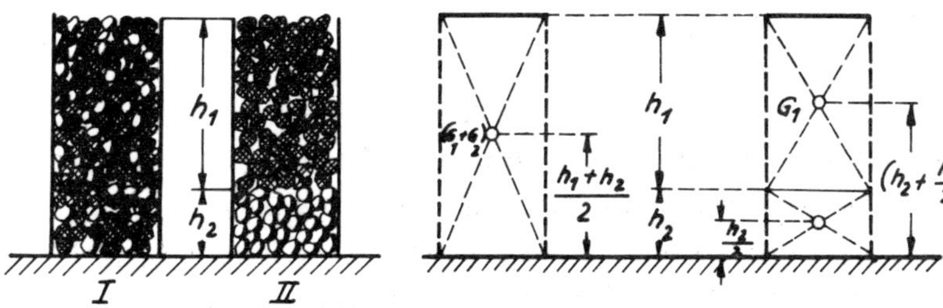

$$\Delta E = \frac{1}{2}(G_2 h_1 - G_1 h_2)$$

Fig. 154

Williams (320), (323) has investigated the subject of size segregation. For a certain vibrating condition a larger particle will be carried to the surface. This can also be regarded as a sieve effect, because smaller particles can more easily fall in the formed hollow spaces or cavities (Fig. 155) (320).

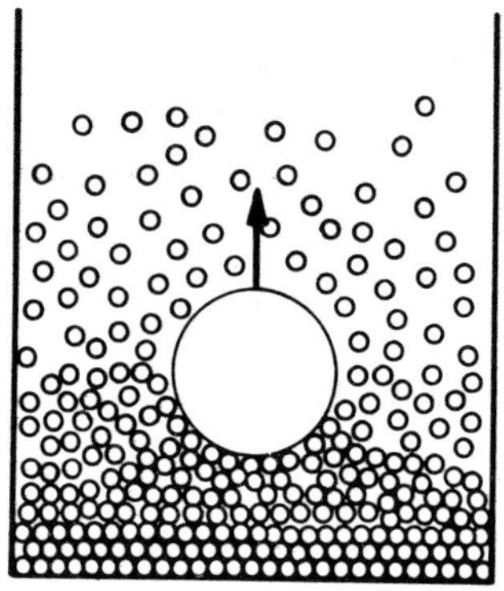

Fig. 155

When applied to a flowing granular bed, this means that smaller and heavier particles are more mobile, which intensifies the segregation.

8.1.2. Degree of Segregation

In order to be able to compare and evaluate different operating conditions, criteria for the degree of segregation must be established. It is sufficient for most cases to compare qualitatively the particle characteristics of the outflowing material. For example, Mitchell (200) showed that by side filling of the bin the variations in size consist are smaller than in the case of central filling and that segregation during discharge is negligible if segregation during the filling has been avoided (Fig. 156).

In order to evaluate the degree of segregation quantitatively some authors have suggested the use of the standard deviation σ as a suitable criterion (82), (249), (267), (289).

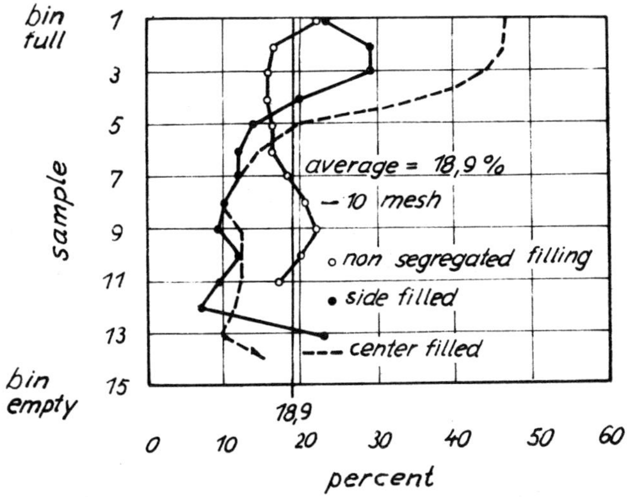

Fig. 156

$$\sigma = \sqrt{\frac{1}{n-1} \sum_{i=1}^{n} (x_i - \bar{x})^2}$$

n = number of samples

x_i = measured value

$\bar{x}$ = desired value

The standard deviation is especially valuable for measurements in plant operations, because only representative samples can be taken there. A much more distinct criterion is defined by Matthée (195), (196). He sums up the differences between the components of each samples and puts this sum in relation to the total stored quantity

$$f_E = \frac{\sum_{i=1}^{n} |\delta_i|}{G} \cdot 100 \ [\%]$$

$\delta_i = x_{1i} - x_{2i}$

x_{1i} = coarse component

x_{2i} = fine component

G = total stored quantity

The segregation factor f_E represents the deviating quantity or the amount of segregation given in percent. This factor is only applicable to model investigations, where exact measurements are possible and the total discharge is analysed.

Determining the parameters that influence segregation, direct and indirect factors are to be considered. Directly influencing factors can be expected from the material characteristics. Based on the results of Fisher (94) who has investigated the case of solid-solid mixing, van Denburg and Bauer (82) have tabulated these factors:

grain size distribution,
nominal grain size,
particle shape,
effective particle density,
surface characteristics, lubricity etc.,
bulk density,
flow ability,
angle of repose,
resistance to agglomeration.

The density distribution must also be added to these factors. Most important of all these factors are the nominal grain size, size distribution and density distribution. Williams (321) found that the tendency to segregate decreases with reduced particle size because the influence of cohesion increases. With sizes less than 20 μ no segregation takes place any more. According to

Matthée, the particle shape is to be regarded as a primary factor, because segregation during filling can only take place when the particles are rolling down the surface of the pile. A few investigations confirm this (82).

Factors indirectly influencing the degree of segregation consist of limiting conditions, such as

bunker diameter D,
orifice diameter d,
slope of hopper wall α ,
filling height H,
wall friction tan δ ,
bunker shape,
auxiliary equipment,
filling location and filling method,
discharge point and discharge method.

8.1.3. Measures to Reduce the Degree of Segregation

Using experimental data and information from literature survey, a list of techniques to reduce segregation are being presented here. It should be kept in mind that a number of these are still to be regarded as hypothetical until further experimental results are obtained. In this connection one must realize that materials which tend to segregate are mostly free flowing. Therefore an eventual discrepancy between ways to reduce segregation and ways to improve the flow is not serious.

The possibilities of minimizing the direct influences through variations of material characteristics are very limited. In accordance with process requirements, the material should be:

crushed to a possibly uniform grain-size (321),

pulverised as fine as possible (321),

stored wet to avoid any segregation through agglomeration (321).

Further means are given by varying bin design and dimensions, specially by using mass-flow bins (132), (133), (249), (267), (268) which limit very effectively the freedom of particle movements during the bedding. Further more the bulk material is discharged in horizontal layers with the same particle composition. The requirement to use high H/D ratios in bin design is also stressed by Matthée (196), because after reaching a certain segregation peak the values for f_E are greatly reduced (Fig. 157).

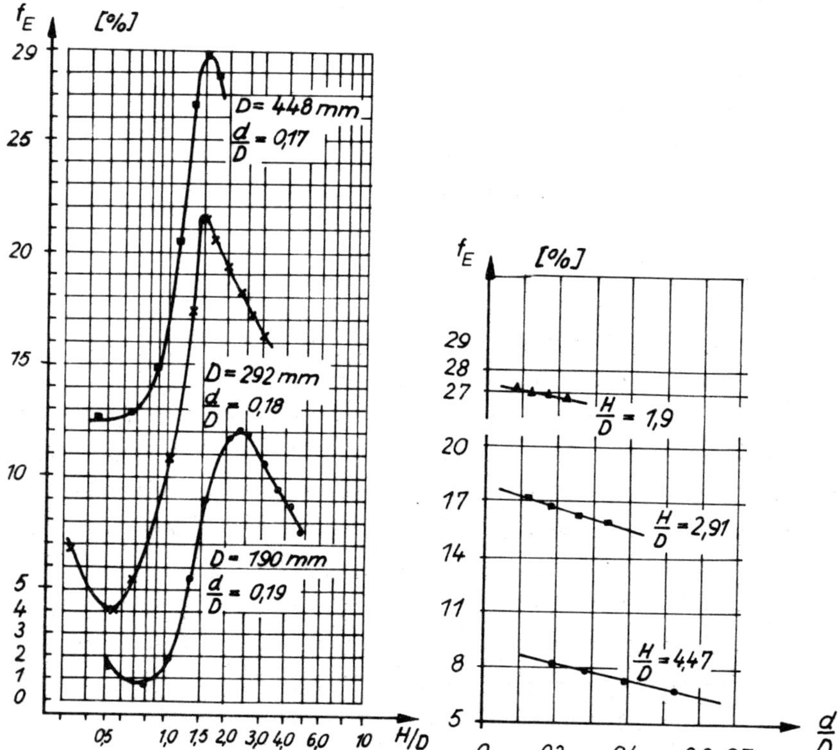

Fig. 157

A segregation minimum occurs further at very small H/D ratios (196). This fact is, at first glance, very surprising and can be explained only by the fact, that a crater at the material surface is formed and thus a remixing occurs very quickly after flow starts during discharging. To a small degree segregation is reduced by widening of the outlet (169) (Fig. 157).

If the material is not piled up during bunker filling and is stored in a non-segregated fashion, then, according to Mitchell (200) and Peacock (223), no appreciable segregation occurs. A series of technical solutions are suggested in the literature, and shown in Fig. 158.

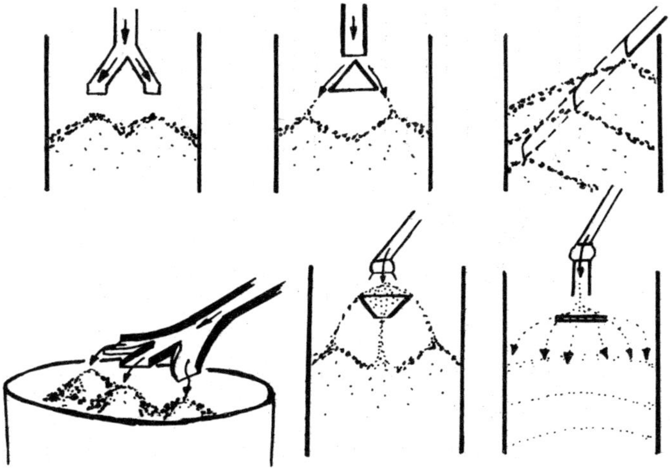

Fig. 158

In most cases, a multiple-point filling is sufficient, because one pile is broken up into a number of small ones. By suitable arrangement of filling points fractions of all components are carried to the central core as well as to wall zones (103), (249), (295).

Other methods provide for a uniform distribution over the whole cross sectional area of the bin (295). Peacock (223) suggests to divide the bunker into a number of small compartments by a system of vertical baffle plates to avoid any possible horizontal movement of individual particles (Fig. 159). According to Richards (249) the freedom of movement of individual particles can be reduced by rapid charging.

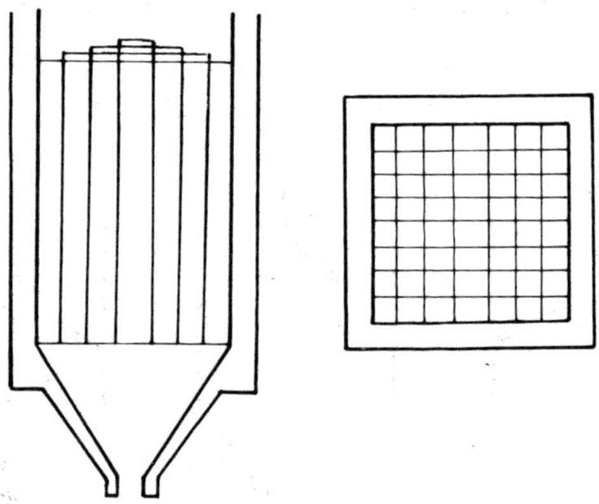

Fig. 159

Further, it is possible to utilize the core-flow effect for a remixing during emptying. Through installation of inserts, the flowing zones can be widened and thus a substantially bigger part of the bin can be activated (73), (82), (103), (200), (295). The discharge from two or more outlets (Fig. 160) can also lead to remixing when coarse and fine materials are equally drawn from the developed flow zones (271), (295). Moreover the surface is divided into a number of small cones. In this connection, it cannot be generally said that long slopes are to be avoided (267). Mitchell (200) (Fig. 156) for example could prove that with side-filling the degree of segregation can be reduced in comparison with center filling. The results of van Denburg and Bauer (82) and Williams (249) are presented in

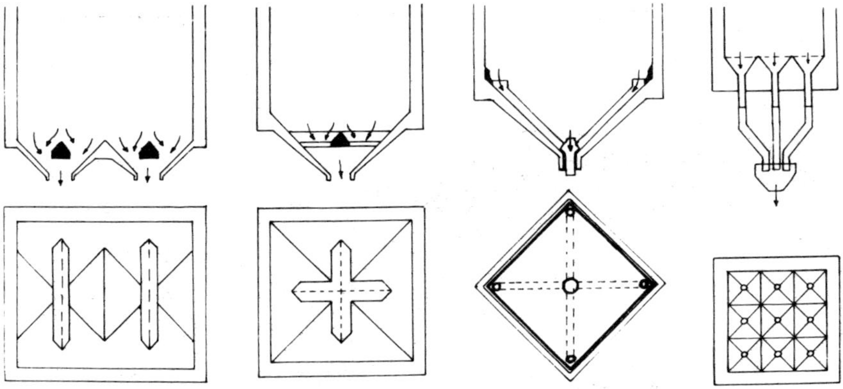

Fig. 160

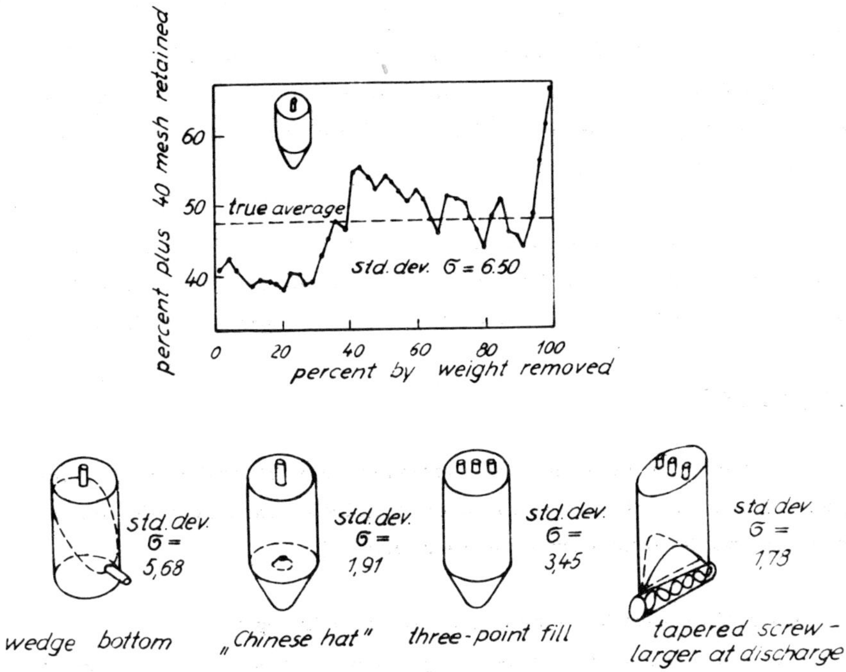

Fig. 161

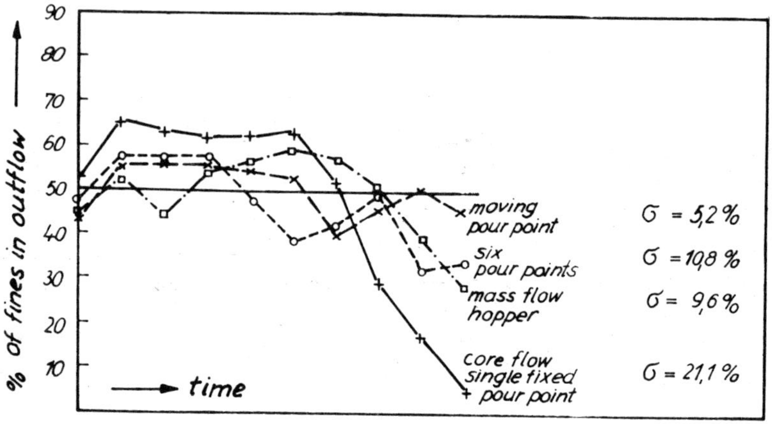

Fig. 162

Figures 161 and 162. The different values of standard deviation are due to the different materials, grain distribution and models tested. They can be marked as the first steps to evaluate methods for controlling or reducing segregation systematically and quantitatively.

The last measure to minimize segregation is the suitable construction of mechanical equipments for discharge, so that the whole cross sectional area of the bunker can be activated and mass flow achieved (see Chapter 6).

8.2. Degradation

In many technical processes, where granular materials with coarse to fine grain sizes are handled, degradation occurs. Depending on the situation there are two different kinds of degradation (254):

a) intentional degradation: blasting, crushing, grinding etc.
b) unintentional degradation: classifying, loading, transport, storing in bins and bunkers etc.

In this chapter, the attention will be confined to the degradation caused by storing a material in bins and bunkers. It is often very important to prevent the degradation because of the further processing of the stored material.

Degradation usually occurs when material is charged into the bin, but also during withdrawal further degradation may occur.

Degradation of the material can be reduced by the following methods:

a) by keeping the bunker filled (185), (247), (248), (328) so that the material does not drop freely over the full height of a bunker causing breakage and also increasing the cohesive strength of fine material,

b) by the material sliding downwards along an inclined wall of the bunker,

c) by using plastic liners (F 45), (F 61) or fine particles to cushion the fall to the lowest point of the bunker,

d) by dropping large charges as a unit load. It has been proven (63) that, for example, in a bunker charged from a car dumper or large skip, the degradation is less than charged by a conveyor belt. That is because of the fact that the inside of the mass is protected by the material on the outside. However, increased packing of the material becomes a greater problem,

e) by installing inserts and special cascade constructions. Some examples are given in the following figures.

Fig. 163 shows three methods which improve the filling of the bunker. Besides reducing degradation, they are also helpful to prevent segregation.

Tests for determining coal breakage in bunkers fitted with cascade chutes (center) (7), (242) have shown that a freely dropped coal had a degradation of 44 %. By using cascade chutes, the amount was reduced to 3, 5 %, and by

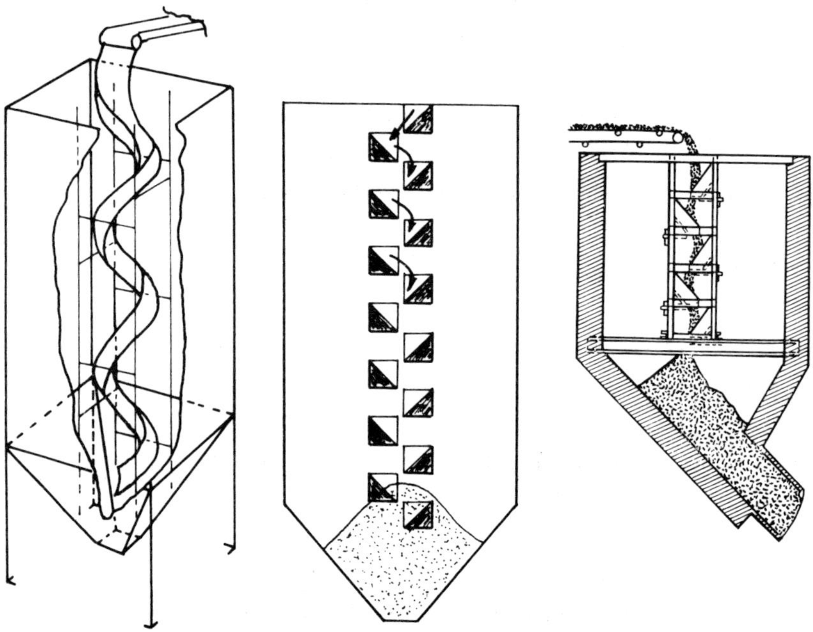

Fig. 163

using spiral chutes (left), the degradation was less than 2 % (301). The third cascade chute (right) is fitted with adjustable inclined steps providing the best flow condition for a certain material (157). This type of chute works also successful in open-air storage (158).

Another method is placing the spiral chute in the wall of a bunker (F 29) (242). The construction is made from pre-cast reinforced concrete units.

The surface is protected by fused cast basalt lining. Fig. 164 shows a typical illustration of a wall section. This method is said to be cheaper than steel

Fig. 164

constructions and does not impair the material flow. Among the developments of the NCB West Midlands Division to reduce coal breakage in bunkers, two more designs should be mentioned (6). Fig. 165 shows the sketch of a bunker with a telescopic chute for loading and conventional discharge. This chute is retractable and reduces the free fall of the material, as the lowest part of the chute is always near the surface of the bulk material. In Fig. 166 a chimney chute and a movable curtain are combined. The curtain, for controlling the flow in the chimney, can be moved up- or downwards so that the coal, which is fed by a belt into the chimney chute, can be gently deposited in the bunker. In comparison to a breakage of 44 % for freely dropped coal, the telescopic chute decreases breakage to about 18, 2 % and the chimney and curtain loading to 3, 3 %.

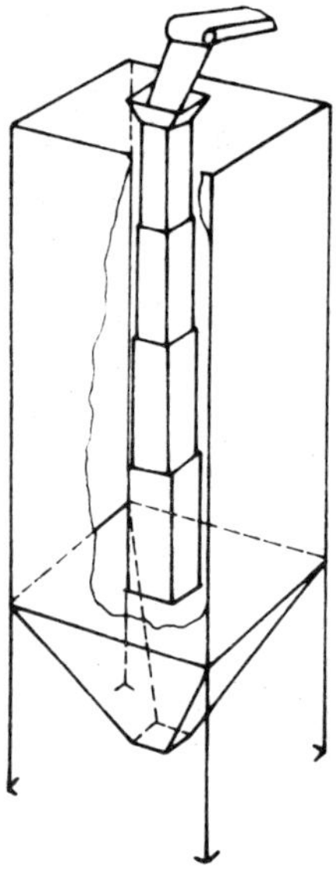

Fig. 165

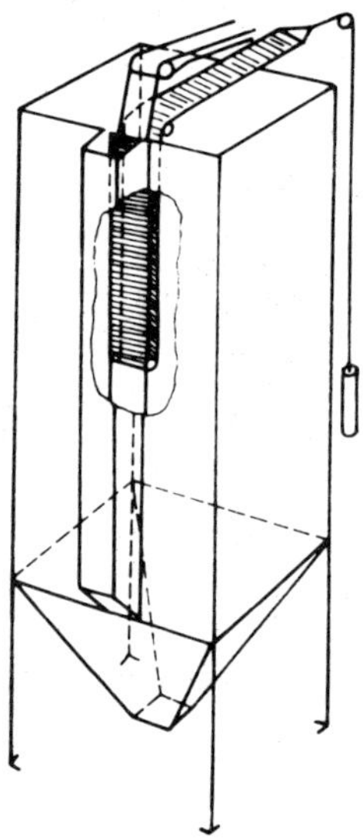

Fig. 166

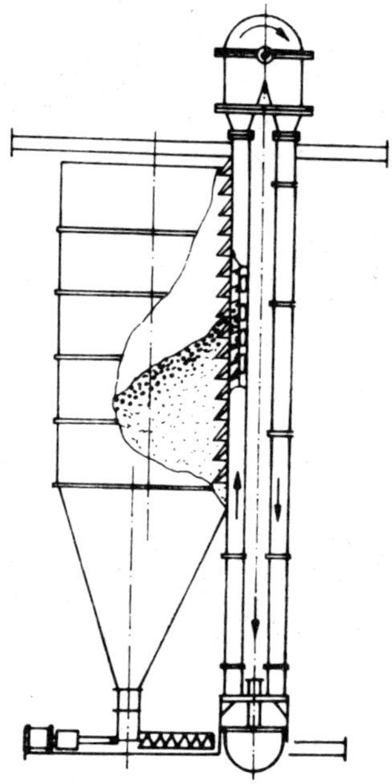

Fig. 167

A fairly new method for filling bunkers and bins is realised by using a drag-chain or flight conveyor (F 8). Fig. 167 illustrates this operation. The material is only lifted as high as required by the existing level of material in the bunker. This eliminates free fall during the whole filling process and the material slides gently and heaps itself according to its own angle of repose. The most important characteristic of bin filling by a flight-conveyor is the stationary louver section, adjacent to the conveyor housing, allowing the material to flow out of the conveyor, but not back into the conveyor housing.

Degradation during discharge can be reduced when mass flow of the material is attained (232). This is possible, as mentioned before, by Jenike´s method (133) of calculating suitable bunker geometry. In cases where this method cannot be applied, an increase of the bunker activity can be achieved by

a) enlarged area of the outlet,
b) steep and polished hopper walls,
c) several outlets in suitable positions,
d) placing inserts above the outlet,
e) chimney discharge.

The items a - d have already been discussed in Chapter 4 and their importance have been shown. The chimney discharge (163) is an effective means of reducing breakage. This chimney is placed in one wall of the bunker either alone (Fig. 168) or in conjunction with spiral chutes or curtains for charging (6). A vertical slot is arranged along the whole length of the bunker and the material feeds gently into the chimney.

This chimney design reduces the degradation predominantly for non-caking lump materials. The chimney may also be fitted inside the bin against one wall or centrally located in the bin.

The cross section of the chimney may also be semi-circular. Kvapil (163) proposes that the radius be enlarged at a rate of 3,5 % towards the outlet, so that the friction between material and inside wall is reduced. Fig. 169 (167) shows a sketch of such a bin with chimney discharge and the position and shape of the outlet. The diagram represents the minimum width of the slots b, the diameter of the chimney d and the minimum height of the discharge opening v in relation to the maximum particle size D.

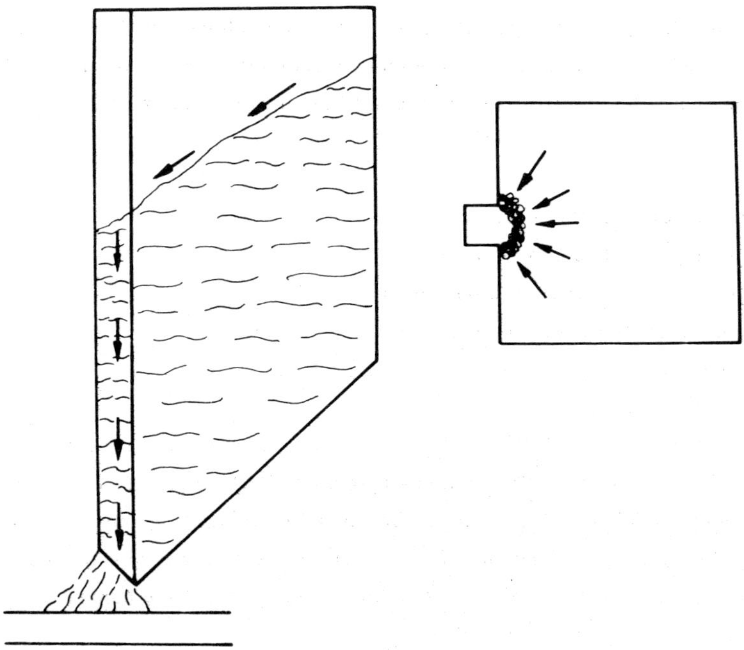

Fig. 168

8. 3. Abrasion

Another important factor to be considered for storing bulk materials in bins is the abrasion which takes place on the walls and bottoms caused by the movement of the material during charging or discharging. This abrasion depends mainly on the hardness and size of the stored materials and on the hopper wall material itself.

The abrasion does not only influence the life of the bins and bunkers but also disturbs the discharging. Due to the friction, an arch of the bulk ma-

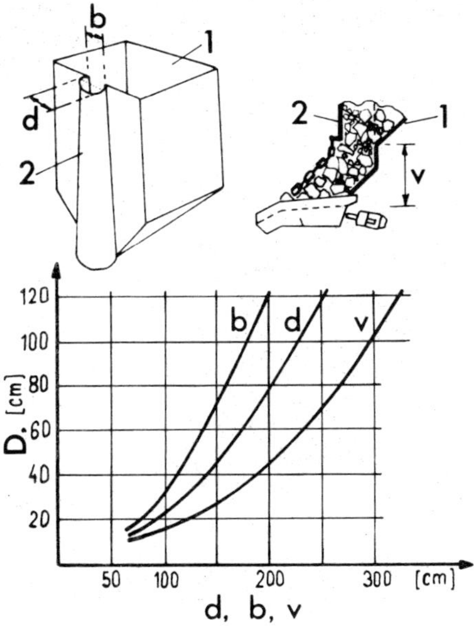

Fig. 169

terial builds up in the interior. The friction conditions between material and wall can be favourably influenced by a smooth lining.

For the storage of abrasive materials, it is necessary to avoid abrasion on the walls. When steel or concrete hoppers are charged with iron-ore, coke, limestone or other materials, one successful method is to use fused cast basalt lining (F 51). The thickness of the lining and the shapes of the tiles are chosen to match the operating conditions. For heavy and large sized materials, 50 mm thick hexagonal shapes are installed (Fig. 170). For light materials like coke, 30 mm thick tiles of square shape are sufficient. Because fused cast basalt linings are not effected by corrosion and are not hygroscopic, sticking of humid material to the walls is avoided. A further advantage is the fact that the surface will be very smooth after a short time

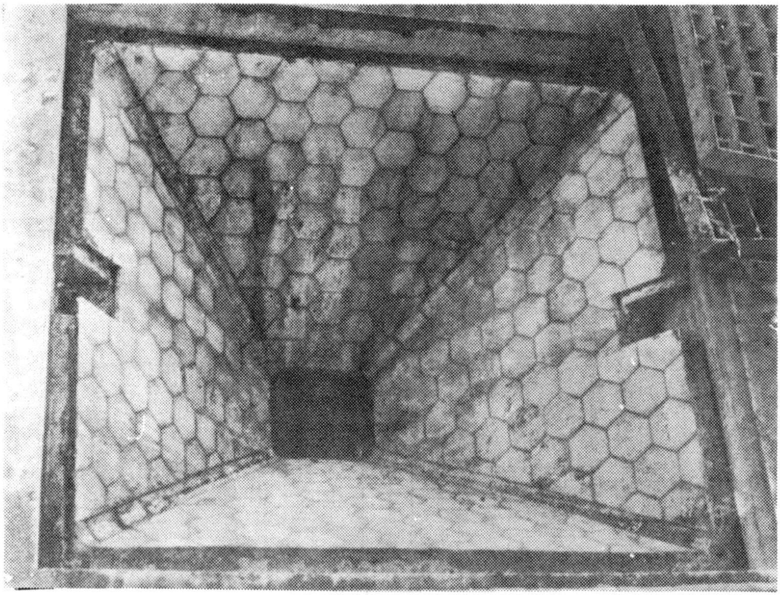

Fig. 170

of service leading to a very reduced coefficient of friction. Other typical examples of special hard tiles or bricks are Accrington red - brick tiles and Haunchwood blue pressed pavers (173).

A fairly new development is the lining with the so called "Lonsica"-concrete using silicon carbide (3), which has been found successful in inclined underground bunkers. After a throughput of 12,000,000 t, no obvious signs of abrasion can be found and the surface is mirror-like.

Gunite, also employed to act as wearing surface, may be applied to the bunker shell by a spraying process. Usually small connectors are affixed to the bunker wall to provide the necessary keying and a light mesh reinforcement for strength. The thickness of the gunite layer is about 2 in. (173). In this case, the application of structural mechanics is of great importance (188). Chrome-alloy cast-iron liners may also constitute a solution (328). Fig. 171 comprises the abrasion resistance of various liner materials. The great advantage of fused cast basalt is clearly to be seen (255).

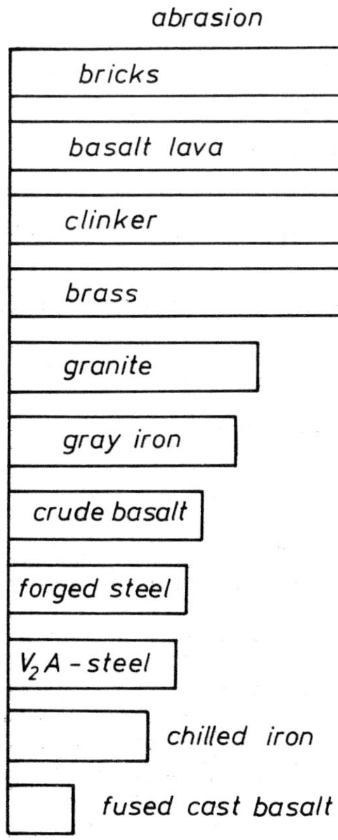

Fig. 171

In the last few years experience was also gained with the use of special rubber linings. Fig. 172 shows the attachment of the rubber sheets which have a high resistance against abrasion and corrosion and a high tensile strength (F 61). As they are only fastened at the top edge, a kind of "fluttering"-motion reduces the possibility of arching or sticking. The use of steelplates with special vulcanized rubber linings is another proven method against abrasion (F 45).

When using plastic linings against abrasion, such as mipolam, vulkollan (93), etc., it is still necessary to find out by experience the thickness of the layer, which mainly depends on the grain size and the dumping height of the stored material. Good results have been achieved by using sheets of a special polyethylene with a high molecular weight (15).

The following are examples for the successfull application of lining bunkers with plastic sheets (125):

a) Iron ore bunker for a sintering plant

grain size	0 - 25 mm
moisture	4 - 7 %
thickness of liners	6 mm

Result: after one year of operation the wear was less than 0.1 mm.

b) Bunker for coking coal

grain size	0 - 6 mm
throughput	3000 t per day

Result: almost no wear at all after five years of operation.

To prevent corrosion of the bin when storing materials many plastic and synthetic resins have been developed. Detailed informations are given in the literature (81), (199), (303).

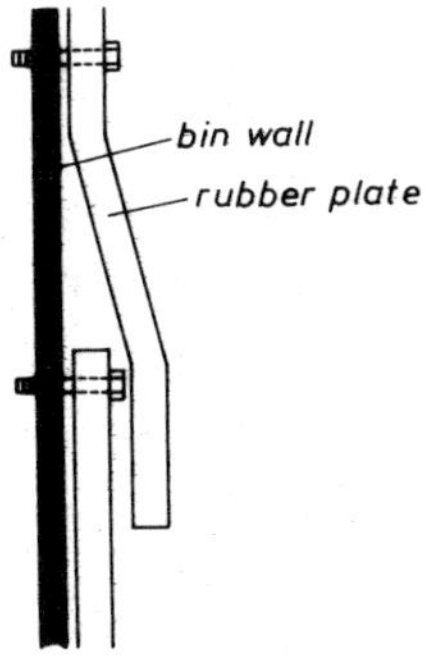

Fig. 172

Russian scientists have experimentally found a method to predict the abrasion of linings in coal bunkers (190). By counting the number of impacts of falling anthracite on different test plates in the hopper and by comparing the corresponding loss of lining thickness, a factor for abrasion was developed. Laboratory tests for the same purpose were run in a rotary drum. The results could be transferred to real conditions in practice.

9. Control and Automatic Equipment

9.1. Level Measurement

A large number of indicators and methods exist for measuring the level of bulk materials stored in bins, bunkers, silos, or other containers. The use of these measuring devices and their arrangement must be carefully adapted to the chemical and physical properties of the stored material and to the local conditions. Generally, there are two basic methods of measurement, depending on what is to be achieved:

a) continuous level measurement,
b) limit-level measurement.

Continuous level measurement is required if the height of the material in the bin must be known at any time. An automatic recording may be provided.

Limit-level measurement serves to ascertain the material level at a certain point, such as maximum or minimum levels. This method is important to regulate feeders, dischargers, pumps etc. which operate in connection with the storage system. Besides the basic purpose of these two methods, the choice of the devices to measure the level of bulk materials is governed by the following factors:

a) local or remote recording of the material level,
b) dimensions, structural materials, and location of the bin, types of feeders and dischargers,
c) grain size, flowability, chemical and physical properties of the stored materials, temperature inside the bulk materials and of ambient temperature,
d) pressure conditions in the bins.

Only when all these factors have been properly taken into consideration, a reliable level measurement and automatic level control may be achieved. Since the service requirements differ considerably from each other, many types of level indicators have been developed to meet the different operating conditions. A complete and detailed description of all of them would be too voluminous for this publication. Therefore, the following discussion only covers some typical applications and gives some recommendations (see Fig. 173), (46), (47), (48), (95), (203). (269), (285).

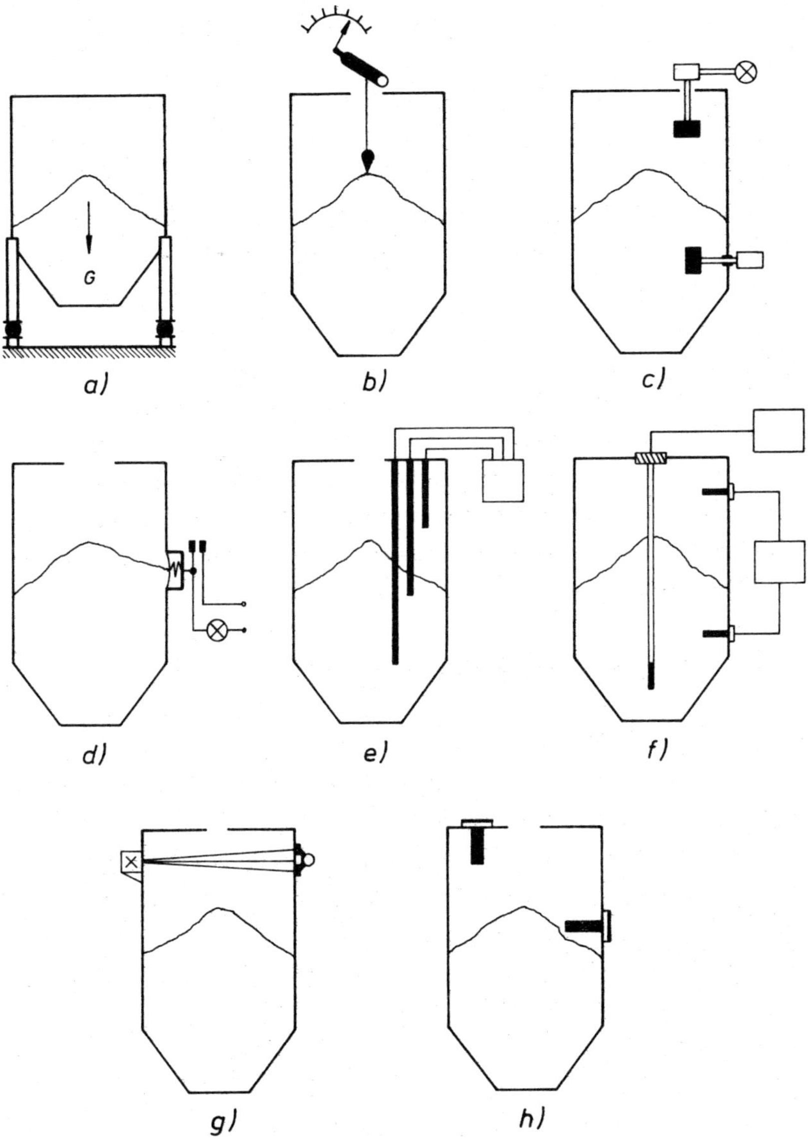

Fig. 173

(The manufacturers listed do not constitute a complete list of firms. This list should only be considered as a guide, similar to the other chapters).

9.1.1. Determination of Material Weight in the Bin

Throughout the years, one of the principal means of determining the quantity of material in bins, hoppers and other containers has been to weigh the content. The weight used to be determined by a standard mechanical weighing arrangement. However, the application of electrical control systems and the development of load cells has replaced the mechanical weighing system (F 18). In case of load cell application the container is directly mounted on the cells, in such a way that the whole weight rests on them (Fig. 173 a). Depending on the shape of the container, three or four cells are normally used. The load on the cell is transduced into an electric signal and the weight can be directly determined. The maximum accuracy is said to be 0.1 % when using four cells for a total load of 10,000 tons (i. e. 2,500 tons on each cell).

9.1.2. Electromechanical Sounding Probes

A plumb bob fixed on the end of a steel wire is lowered by a power driven winch to the level of bulk material (Fig. 173 b). When the material is reached, the wire tension is reduced generating a signal which records the material height in the bin. As this operation may be periodically repeated, a continuous level measurement is possible for all grain sizes. The installation is suitable for bins with a depth up to ~ 30 m and the accuracy of measurement is about $\pm$ 2 cm (F 20), (F 57), (2).

9.1.3. Rotating Paddle

Motor driven paddles rotate as long as they are not blocked by the bulk materials. As soon as rotation stops, a signal indicates the specific limit level depending on the position of the devices. They can be mounted as shown in Fig. 173 c and for a vertical position the shaft can be longer than required for a horizontal position. Their application is mainly recommended for materials with good flowability, predominantly for fine grains and powders and an accuracy of $\pm$ 2 cm in level height can be expected (26), (46), (F 5), (F 13), (F 38).

9.1.4. Diaphragm-Type Level Control

This type was one of the first generally used for limit-level measurement in the industry. The unit is easily installed in the wall of the bin, and the system utilizes a pressure-sensitive diaphragm-operated switch (Fig. 173 d). The diaphragm, made from stainless steel, neoprene, rubber or teflon, has a limited flexibility and a restricted travel. It is designed for sensing the pressure exerted by dry, free flowing bulk materials. The switch serves to actuate visual or audible signals as well as bin filling and discharge controls (F 5).

9.1.5. Electrical-Resistance Sensor

This fairly new continuous level sensing element is suspended like a wire from the roof of the bin into the hopper. It consists of a precision-wound electrical-resistance helix. A composite jacket serves to enclose and protect the inner electrical system and acts as a diaphragm which receives the pressure of the surrounding material, causing a contact between the resistance element and the conducting base strip. By special recording devices, the location of the material-air interface, i. e. the level of material, can be determined. Greatest experience with these level indicators has been with plastic pellets in metal silos. However, sensors for heavy minerals with tensile strengths in excess of three tons have also been developed (86), (F 41).

9.1.6. Level Control by Conductivity

With this technique of limit-level measurement, the transducer consists of a partly insulated electrode or probe. This electrode, in varying lengths, is installed inside the container and connected to a controller (Fig. 173 e). The wall of the container forms the other pole of the conductive circuit, or, with non-metallic containers a suitable "contra electrode" must be fitted. The level of the stored material, which must have electrical conductivity, is allowed to rise until it touches the end of the electrode. This then closes the current and the controller starts signalling (F 13), (F 20).

9.1.7. Level Control by Capacitance

Capacitance level indicators need a measuring element or electrode to be installed inside the bunker. This forms one part of an electrical capacitator, and the change of the electrical capacitance measurable between this electrode and the container wall provide the basis for the level indication (300). The electrodes may be of the rod or rope-type. Depending on the method of installation, the sensing elements may be arranged so that they measure either a continuous or limit level (Fig. 173 f). Long cable probes should be mounted near the center of cylindrical silos, because in this position a lateral movement of the probe, due to material flow is prevented thus minimizing measuring errors. Furthermore, the probe should never reach down to the hopper outlet but terminate at about the middle of the cone. Long probes may be subjected to rather high tensile forces when the bin is emptied. While the peak value of such forces may last only a fraction of a second, the wire as well as the probe supports must be strong enough to withstand them (Fig. 174). Detailed information on installation, insulation of the probes and many application examples are available in different languages (F 20).

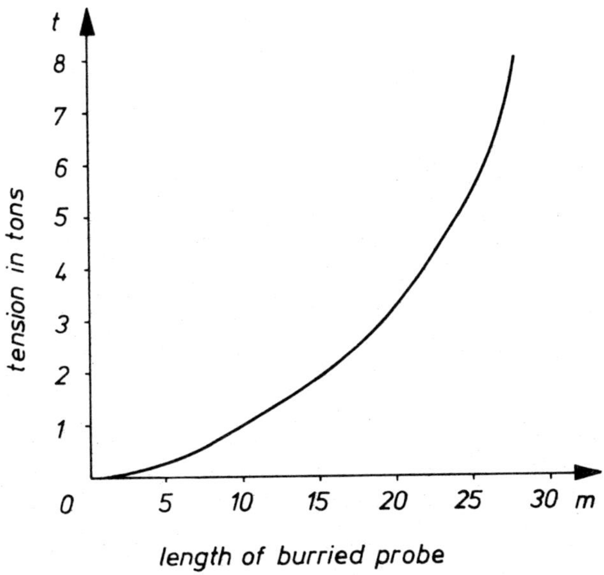

Fig. 174

9.1.8. Level Control with Radioactive Isotopes

In cases where level-measuring devices cannot be mounted in the inside of a bin, isotopes level controllers are suitable alternatives. Highly corrosive or abrasive bulk materials, extremely high temperatures, or excessive grain size as found in the coal and quarry industries, may make the level measurement by the previously described methods either extremely difficult or impossible. There has been a considerable experience with the radioactive type of level measurement in food and paint industries, mines, cement plants, foundries etc. The technical principle is based on the absorption of the radioactive rays by the stored material. The radioactive source is mounted so that the gamma ray beam is directed crosswise through the bin towards the detector, which transforms the signals into electrical impulses (Fig. 173 g). If the source and the detector is moved vertically along the wall of the bin, a continuous level measurement may be achieved (62), (F 2), (F 20).

9.1.9. Level Control by Ultrasonics

Level measurement of an extremely light and fluffy material with a large amount of entrapped air cannot be easily performed with the methods mentioned above. Fortunately, most of these materials are good sound absorbers. Hence, ultrasonic waves can be applied to produce a continuous and limit level control.

The basic system consists of one or more sensors mounted to the bin as shown in Fig. 173 h. The operation is based on the principle of sonic wave propagation in air and used to measure the air space in the bin between the sensor and the stored material. The accuracy is about ± 1 to 2 cm.

Each method for level measuring has its inherent advantages and disadvantages and every system has its limits. Unfortunately there is no universal instrument that can meet all requirements. The following Table 15 establishes some relationship between material characteristics and measuring principles for continuous level measurements and suggests a selection guide (F 20). If all items of material characteristics are checked with "yes", the suitable measuring principle is found. If an item is questionable or listed at the right column, the choice is led to the next measuring method. In case of limit control only, similar recommendations are suitable (F 20).

Table 15

Material Characteristics	Measuring Principles	
A.	capacitance	
temperature	$< 400^{o}$ C	$> 400^{o}$ C
pressure	$<$ 300 atm.	$>$ 300 atm.
density	stabilized	changes
moisture	stabilized	changes
physical condition	permanent	changes
build-up on the electrode	none	strong
granulation of bulk material	constant	changes
grain size	$<$ 50 mm	$>$ 50 mm
abrasion	slight	strong
	yes	no
B.	electromechanic	
temperature	$< 200^{o}$ C	$> 200^{o}$ C
pressure	0 atm.	$>$ 0 atm.
	yes	no
C.	ultrasonics	
temperature	$- 20^{o}$ C up to $+ 60^{o}$ C	$< 20^{o}$ C $> 60^{o}$ C
pressure	0 atm.	$>$ 0 atm.
sound reflection	good	bad
high air movements	none	yes
	yes	no
D.	radioactive	
calculate specific weight (g/cm^3) x distances between detector and source (cm)	$>$ 22.5 g/cm^2	$<$ 22.5 g/cm^2
	yes	no

special consulting necessary

9.2. Flow Measurement

Devices for determining the rate of material discharge from a bin are often an important part of the total system. Measuring or controlling the flow rate is done by means of volumetric, gravimetric or other methods. The required accuracy depends on the succeeding technological processes in a plant. For instance, in crushing plants a variation of ± 10 % of the mean rate would be acceptable, while in some chemical processes an accuracy of better than ± 5 % is required. Generally speaking, higher precision control will be required the closer a storing system is to the final steps of a technological process (249).

9.2.1. Volumetric Feeders

Volumetric feeders deliver a given volume of material per unit time, and its discharge rate is determined by multiplying the volume with the bulk density of the material being fed. As the bulk density varies the discharge rate varies too.

The different types of volumetric feeders and their range of application are discussed in the Chapter 6.1. on mechanical discharge arrangements. Volumetric feeder control may be accomplished by:

a) frequency and amplitude for vibratory feeders,
b) diameter and rotation speed for plow feeders, table feeders and screw feeders,
c) velocity of belt feeders, apron feeders and scraper feeders,
d) rotation speed and size of pockets of vane feeders.

9.2.2. Gravimetric Feeders

If an accurate feed rate control is necessary and the material has variable physical characteristics, a weighing device should be used in connection with feeders. In case of a gravimetric feeder, the material is weighed when passing a fixed point on a conveyor belt. One method of weighing on a conveyor belt requires that a part of the belt is supported by a pivoted arm (35), (152), (Fig. 175).

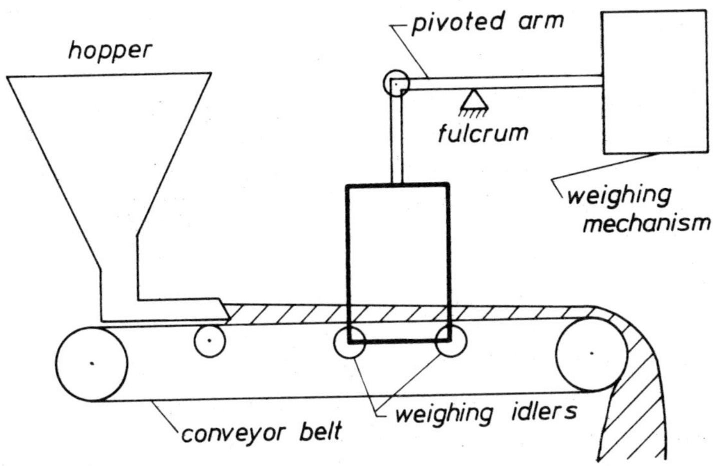

Fig. 175

The weighing mechanism counterbalances the arm and the restoring force is thus proportional to the weight of the material on the belt. More modern instruments are electronic belt scales. The load of the material on the belt acts on idlers which are mounted on load cells, and the output from the load cells must be proportional to the weight of the solid. By multiplying with the speed of the belt the mass flow is obtained (77), (231), (304).

Typical arrangements for weighing units using load cells, are shown in Fig. 176, as described by B a l d w i n (part of (249)).

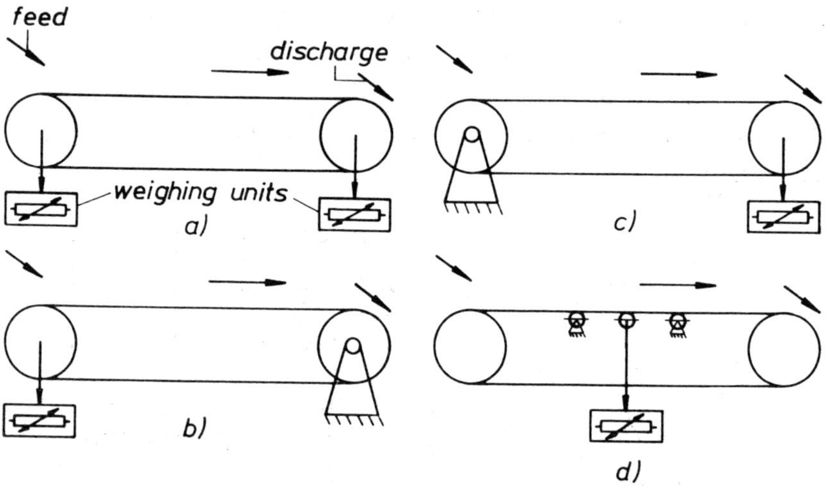

Fig. 176

a) the entire conveyor is placed on weigh scales,

b) the conveyor frame is designed as a cantilever with the pivot at the discharge end and the weighing cell at the feed end,

c) reverse of scheme b: pivot at the feed end and weigh cell at the discharge end,

d) a weighing idler is built into a conveyor and mounted between fixed idlers, as mentioned above.

An error in weighing may occur due to the impact effect on the scale if the material falls from a certain height. Therefore the fall height should be kept low, or in scheme c the material falls over the pivot, or in scheme d the impact does not affect the weighing idler. Another error may occur if the cohesive nature of the material varies. At the discharge end of the belt the material may drop off readily, leaving a low angle of repose and less material to be weighed, but in the other extreme the material may protrude beyond the drum and more material will be weighed. These effects would be more pronounced in scheme c, while scheme d seems to be the least affected. However, scheme d has one disadvantage: it weighs the material through the belt and belt tension might have an effect. Using relatively short belts and fairly dense materials (about 60 lg/ft^3) this effect will not be too great. For lighter materials the weigh belts must be thin and flexible and this method might not be suitable. A solution has been found by using a platform supporting the belt along its full load carrying length with a section in the middle used for weighing.

If a constant weight of the stream of discharged material must be maintained the feeders are designed with arrangements which allow automatic control for metering.

Only a few examples are given as illustration, since there are too many types which could be described here (F 4), (F 6), (F 40), (F 49), (F 50).

Fig. 177 shows the sketch of such a feeder which is pivoted to the control gate. It operates by a counterweight that controls the opening of the gate by the weight of material on the belt (181) (F 27).

Fig. 178 shows examples of combination of discharge units with a weigh belt. As soon as the weight of material on the belt deviates from the exact value the mechanical discharge arrangements are controlled automatically so that the discharge rate may be changed (F 1), (F 19), (F 58):

a) weigh belt with vibratory feeder controlled by the amplitude,
b) weigh belt with belt feeder, controlled by the speed,
c) weigh belt with screw feeder, controlled by the number of revolutions,

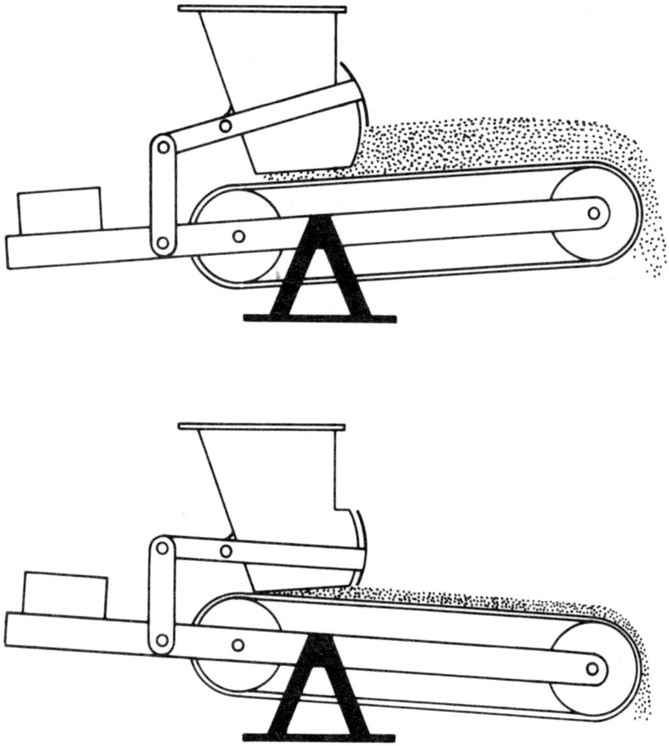

Fig. 177

d) weigh apron conveyor with a pneumatically operated and controlled gate.

9. 2. 3. Special Methods

In the following paragraphs some new methods of solids flow measurement are mentioned including some information on their range of application. Details of their operating principles and structural designs are given in the literature (35).

9.2.3.1. Optical Method

This method is volumetric and can only be used with materials having a high coefficient of reflection, such as a detergent powder being transported on a conveyor. It makes no allowance for variations in the packing density of the material and its accuracy can be seriously affected by dust settling on the optical surfaces. The fundamental idea is that a beam of light is projected on to the material on the conveyor, and an image of the cross-section of the material is projected on to a photocell.

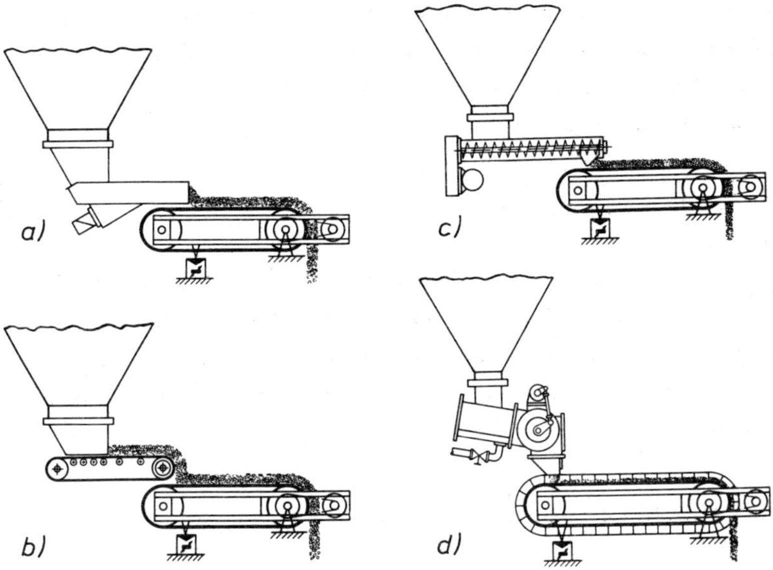

Fig. 178

9. 2. 3. 2. Nuclear Method

This type of flow measurement can be successfully used with belt conveyors, vibrating conveyors, and chain conveyors for cement, grain etc., where the belt velocity can be readily measured. This method is related to the principle, that the amount of gamma radiation absorbed by a material depends on the mass of material in the path of the radiation. Variations in moisture content and density may introduce errors.

9. 2. 3. 3. Rotating Impeller

Rotating impeller flowmeters can only be used in ducts where the solid material is falling freely under gravity. The material falls on to a rotating impeller which flings the solid outwards. A measurable torque is produced on the impeller motor shaft that is proportional to the mass flow rate of the solid. This flowmeter is suitable for free flowing fine grains and powders which should not be too abrasive.

9. 2. 3. 4. Spiral Vane Flowmeter

Quite different from the methods mentioned above, this instrument is a spiral vane flowmeter. It is used in vertical ducts where the flowing solid revolves the spiral vane. The number of revolutions is related to the volumetric flow rate. The limitations of application are similar to those of the rotating flowmeter.

9. 2. 3. 5. Inclined Plate Method

The material falls onto an inclined plate attached to a rigid beam which is pivoted about a point near its center. The moment is measured by a force balance unit. If the velocity of the material is constant the force on the plate is proportional to the mass flow rate. It has been successfully applied even for coke and iron ore, but the solids should be densely flowing. Because of wear necessary maintenance may be considerable. This method can also be used in combination with electronic control (68).

9. 2. 3. 6. Method of Heat Labeling

A heated layer is introduced into the stream of granular material in which two groups of thermocouples are placed at certain distances apart. These give impulses to start and stop an electric stop watch. Experiments have been performed in cylindrical vessels, 300 mm wide and 1400 mm high, with quartz sand of less than 1 mm grain size (107).

References

1. Anon. : Bin avoids segregation and handles difficult material. World Mining 22 (1969), No. 3, p. 68.

2. Anon. : Bunker level control at Wolstanton. Colliery Guardian (1970), No. 2, pp. 79 - 80.

3. Anon. : Coal bunkers with abrasion resistant concrete lining (Rohkohlenbunker mit verschleißfester Beton-Auskleidung) (German). Informations of Steinkohlenbergbauverein Essen, Germany, 1970, Publ.-No. 76, p. 3.

4. Anon. : The effect of the use of an underground bunker on the size consist of the saleable coal at Linby Colliery. Issued by the East Midlands Division, No. 6 Area Scientific Department, National Coal Board, Great Britain, 1958.

5. Anon. : The degradation of coal in underground bunkers at Ormond Colliery. Issued by the East Midlands Division, No. 6 Area Scientific Department, National Coal Board, Great Britain, 1958.

6. Anon. : Bunker design to reduce coal degradation. Issued by the West Midlands Division, National Coal Board, Great Britain, 1964.

7. Anon. : Tests to measure coal breakage in bunkers fitted with cascade conveyors. National Coal Board, Great Britain, Report No. C. D. 8.

8. Anon. : Standard DIN 1055: Design loads for buildings; page 6: loads in silo bins (Lastannahmen für Bauten; Blatt 6: Lasten in Silozellen) (German). Berlin-Köln: Beuth-Vertrieb 1964.

9. Anon. : Standards for construction of silos for bulk materials (Russian). Moscow: Isdatelstvo Stroitelnj Literatur 1965.

10. Anon. : Bunkers and silos for storage of bulk solids (Bunker und Silos zur Speicherung von Schüttgut) (German). VDI-Instruction No. 2694; Düsseldorf: VDI-Verlag 1970.

11. Anon. : Level control in silos (Füllstandskontrolle von Siloanlagen) (German). Die Mühle 100 (1963), No. 9.

12. Anon. : Course on flow and storage of bulk solids. Univ. Bradford, England, March 1970.

13. Anon. : Difficulties in the discharge of coal from bunkers (Nachrutschschwierigkeiten in Kohlebunkern) (German). Energie 9 (1957), No. 3, pp. 99 - 100.

14. Anon. : Gates and feeders for bins (Bunkerverschlüsse und Beschickungseinrichtungen) (German). Hebezeuge und Fördermittel 2 (1962), pp. 380 - 381.

15. Achtelik, H. ; Rado, K. : How to improve material sliding on bunker walls (Verbesserung der Rutschfähigkeit von Bunkerwänden) (German). Information of Ruhrkohlenberatung, Essen, Germany.

16. Alferow, K. W. : Bunkers, locking devices and feeders (Russian). Moscow: Maschgis 1946.

17. Allen, D. S. : Bulk material flow through hoppers and feeders. Chem. Engng. Progress 62 (1966), No. 11, pp. 65 - 68.

18. Andrews, C. K. : The performance of helical screw equipment for the handling of solids. ASME-Publ., No. 68-MH-38.

19. Andrews, C. K. : Bunker design for effective vibration. Symposium "Practical Bunker Design", IMechE, London, 25th. February 1969, paper 5.

20. Aoki, R. et al. : Bins and discharge arrangements for the storage of fine-grained materials (Japanese). Tokyo: Nikkankogyo-Shinbunsha 1963.

21. Ashton, M. D. ; Farley, R. ; Valentin, F. H. H. : An improved apparatus for measuring the tensile strength of powders. J. Sci. Instrum. 41 (1964), pp. 763 - 765.

22. Ashton, M. D. ; Cheng, D. C. -H. ; Farley, R. ; Valentin, F. H. H. : Some investigations into the strength and flow properties of powders. Rheol. Acta 4 (1965), pp. 206 - 218.

23. Athey, J. D. ; Cutress, J. O. ; Pulfer, R. F. : X-ray investigations of flowing powders. Chem. Engng. Sci. 21 (1966), pp. 835 - 836.

24. Aumund, H. ; Mechtold, F. : Vertical and horizontal transport systems (Hebe- und Förderanlagen) (German). Berlin, Heidelberg, New York: Springer Verlag 1969.

25. Aytaman, V. : Causes of hanging in ore chutes. Canad. Mining J. (1960), Nov., pp. 77 - 81.

26. Bahnsen, D.: Electromechanical level measurement (Elektromechanische Füllstandsanzeiger) (German). Aufber. Techn. 10 (1969), pp. 599 - 603.

27. Banks, K.: Practical bunker design for fertilizers. Symposium "Practical Bunker Design", IMechE, London, 25th February 1969, paper 10.

28. Barre, H. J.: Flow of bulk granular materials. Agricultural Engng. (1958), pp. 534 - 539.

29. Bass, M. S.: Multiple-pitch screw equalizes discharge. Chem. Engng. 11 (1960), p. 126.

30. Bates, L.: Entrainment patterns of screw hopper dischargers. ASME-Publ., No. 68-MH-2.

31. Bates, L.: Notes on the selection of hopper discharge screws. Symposium "Practical Bunker Design", IMechE, London, 25th February 1969, paper 11.

32. Bates, L.: Hopper wall profiles. Symposium "Practical Bunker Design", IMechE, London, 25th February 1969, Paper 12.

33. Bates, L.: Handling of bulk solids by helical screw equipment. Brit. Chem. Engng. 14 (1969), pp. 1072 - 1076.

34. Bauer, W. G.: Factors affecting the movement of material through bins, gates and feeders. Pit and Quarry (1945), Aug., pp. 78 - 83.

35. Beck, M. S.; Wainwright, N.: Current industrial methods of solids flow detection and measurement. Powder Technol. 2 (1968/69), pp. 189 - 197.

36. Bedenig, D.; v. d. Decken, C. B.; Hündorf, B.; Rausch, W.: Experimental methods for the direct investigation of the flow pattern of pebble beds (Experimentelle Methoden zur direkten Untersuchung von Fließvorgängen in Kugelschüttungen) (German). Techn. Mitt. Krupp 24 (1966), pp. 147 - 157.

37. Bedenig, D.: Investigations on pressure distribution within pebble beds. Powder Technol. 1 (1967/68), pp. 207 - 212.

38. Bedenig, D.; v. d. Decken, C. B.; Rausch, W.: Investigations on the movement of fuel elements in the core of a pebble bed reactor. ASME-Publ., No. 68-MH-15.

39. Bedenig, D.: A theoretical model for calculating the flow behaviour of a pebble bed and its application to the core of the THTR-prototype. THTR-symposium "Problems of the Pebble Bed and Granular Materials", Jülich, Germany, March 1968, EURATOM-Publ., No. EUR 4190 d, e.

40. Bedenig, D.; Merten, G.: The measurement of pressure distribution in a pebble bed by means of rod-shaped probes. THTR-symposium "Problems of the Pebble Bed and Granular Materials", Jülich, Germany, March 1968, EURATOM-Publ., No. EUR 4190 d, e.

41. Benthaus, F.: Design of bunkers for waste materials and their importance to improve the operation in steep inclined seams (Die Gestaltung von Bergebunkern und deren Bedeutung für die Verbesserung des Betriebsablaufs in der steilen Lagerung) (German). Glückauf 97 (1961), pp. 1061 - 1076.

42. Bernache, P. L.: Flow of dry bulk solids on bin walls. ASME-Publ., No. 68-MH-16.

43. Bernache, P. L.: Flow of dry bulk solids on bin walls. J. Engng. Ind., Trans. ASME (1969), May, pp. 489 - 496.

44. Bessing, R.: The design of mill hoppers in the limestone industry (Die Gestaltung von Mühlenbunkern in der Kalkindustrie) (German). Zement-Kalk-Gips 17 (1964), pp. 15 - 18.

45. Bhattacharya, D.: Direct measurement of the yield strengths of powders. Ph. D. Thesis, Univ. Bradford, England, 1969.

46. Böhme, J.: Technique of level measurement (Füllstandmeßtechnik) (German). Messen und Prüfen (1964), Oct., No. 1.

47. Böhme, J.: Measuring and controlling of container levels in preparation plants (Messen, Steuern und Regeln von Behälterfüllständen in der Aufbereitungstechnik) (German). Aufber. Techn. 6 (1965), pp. 697 - 701.

48. Böhme, J.: Electronic level measurement for containers and silos in mill plants (Elektronische Füllstandmeßgeräte für Behälter und Silos in Mühlen) (German). Die Mühle (1969), pp. 712 - 714.

49. Bohuslav, K. ; Stoyan, D. : A method for dimensioning bunkers in belt-operated mines and its application to the study of broken-belt conveying - belt/bunker/belt (Eine Methode zur Dimensionierung von Bunkern in Bandbetrieben und ihre Anwendung bei der Untersuchung der gebrochenen Bandförderung - Band/Bunker/Band) (German). Bergbautechnik 17 (1967), pp. 368 - 374.

50. Bond, D. C. : Some aspects of filling and withdrawing limestone products from various types of bunkers. Symposium "Practical Bunker Design", IMechE, London, 25th February 1969, paper 1.

51. Brandter, F. : The screw conveyor in its versatile application (Die Förderschnecke in ihrer vielseitigen Anwendung) (German). Aufber. Techn. 10 (169), pp. 424 - 437.

52. Bridal, A. R. ; Miles, J. E. P. ; Schofield, C. : Improving hopper discharge by aeration. Symposium "Practical Bunker Design", IMechE, London, 25th February 1969, paper 19.

53. Brocklesby, M. : Haulage and handling. Colliery Guardian (1965), June, pp. 67 - 70.

54. Brötz, W. : The fundamental principles of a fluidized bed (Grundlagen der Wirbelschichtverfahren) (German). Chem. Ing. Techn. 24 (1952), pp. 60 - 81.

55. Brown, R. L. : The fundamental principles of segregation. J. Inst. Fuel 12 (1939), Oct., pp. 15 - 19.

56. Brown, R. L. ; Hawksley, P. G. W. : The internal flow of granular masses. Fuel 26 (1947), pp. 159 - 173.

57. Brown, R. L. ; Richards, J. C. : Exploratory study of the flow of granules through apertures. Trans. Inst. Chem. Engrs. 37 (1959), pp. 108 - 119.

58. Brown, R. L. ; Richards, J. C. : Profile of flow of granules through apertures. Trans. Inst. Chem. Engrs. 38 (1960), pp. 243 - 256.

59. Brown, R. L. : Minimum energy theorem for flow of dry granules through apertures. Nature 191 (1961), No. 4787, pp. 458 - 461.

60. Brown, J. S. ; Hohmann, E. S. : BCR Easy-flo hoppers let coal flow free at Elrama. Power Engng. (1962), Sept., pp. 55 - 56.

61. Bruff, W.; Jenike, A. W.: A silo for ground anthracite. Powder Technol. 1 (1967/68), pp. 252 - 256.

62. Burbach, W.: Continuous level measurement with radioactive isotopes (Kontinuierliche Füllstandsmessung mit radioaktiven Isotopen) (German). AEG-Mitt. 53 (1963), No. 5/6, pp. 169 - 172.

63. Cameron, D. E.: Coal handling at ports. J. Inst. Fuel 7 (1934), pp. 195 - 216.

64. Carr, R. L.: Evaluating flow properties of solids. Chem. Engng. (1965), Jan., pp. 163 - 167.

65. Carr, R. L.: Classifying flow properties of solids. Chem. Engng. (1965), Febr., pp. 69 - 72.

66. Carr, R. L.: Particle behaviour, storage, and flow. ASME-Publ., No. 68-MH-6.

67. Carroll, P. J.: Hopper designs with vibratory feeders. Chem. Engng. Progress 66 (1970), No. 6, pp. 44 - 49.

68. Chodakowski, J.: Electronic inclined plate mass flowmeter for granular materials. Powder handling and storage, paper 11, Institute of Materials Handling, Maidenhead, Berksh., 1969, E.: A. S. Goldberg.

69. Clark, P.: Volumetric batching of aerated materials. Powder handling and storage, paper 6, Institute of Materials Handling, Maidenhead, Berksh., 1969, Ed.: A. S. Goldberg.

70. Colijn, H.; Carroll, P. J.: Design criteria for bin feeders. AIME Trans. Soc. Min. Engrs. (1968), No. 241.

71. Colijn, H.; Carroll, P. J.: Feeders for bins and hoppers. Brit. Chem. Engng. 15 (1970), Aug.

72. Colijn, H.; Hanson, P. D.: Practical application of hopper and bin design. Canad. Mining and Metallurgical Bull. 73 (1970), July, pp. 180 - 187.

73. Colley, C.: Too coarse, too fine. Mech. Engng. (1963), May, pp. 52-54.

74. Cook, A.: Flow characteristics of bulk materials. Feed Production Handbook. Feed Production School Inc., 1961, pp. 266 - 271.

75. Cooper, F. D.; Garvey, J. R.: Flow of coal in bins. Combustion (1957), Nov., pp. 45 - 49.

76. Cooper, H.: Bunker conveyors in the West Midlands. Iron and Coal 183 (1961), No. 4871, pp. 1101 - 1104.

77. Craven, G. F.: Automatic weighing and batching of bulk solids. Chem. Process Engng. 45 (1964), p. 125.

78. Curland, B.: The puls-jet system for the discharge of powdered bulk materials, in particular flour, from silos (Das Puls-Jet-System zur Austragung von pulverförmigen Schüttgütern, insbesondere Mehl, aus Silozellen) (German). Aufber. Techn. 10 (1969), pp. 557 - 560.

79. Cutress, J. O.; Pulfer, R. F.: X-ray investigations of flowing powders. Powder Technol. 1 (1967/68), pp. 213 - 220.

80. Daeves, K.; Flachsbart, O.: Movement of crowds of people (Strömung von Menschenmengen) (German). VDI-Z. 94 (152), pp. 880-881.

81. Davis, H. E.: Protect concrete and steel. Plant Engng. (1961), pp. 61-70.

82. van Denburg, J. F.; Bauer, W. C.: Segregation of particles in the storage of materials. Chem. Engng. (1964), Sept., pp. 135 - 142.

83. Dubach, P.: New findings in the problem of arching in bin outlets (Neue Erkenntnisse zum Problem der Brückenbildung in Bunkerausläufen) (German). Aufber. Techn. 6 (1965), pp. 50 - 56.

84. Dubach, P.: New findings in the problem of arching in bin outlets (Neue Erkenntnisse zum Problem der Brückenbildung in Bunkerausläufen) (German). Bergbauwissenschaften 13 (1966), pp. 330 - 333.

85. Dumbaugh, G. D.: Concepts of vibratory hoppers. ASME-Publ., No. 65-MH-11.

86. Ehrenfried, A. D.: Level measurement of dry solids with new extended resistive sensor. ASME-Publ., No. 68-MH-37.

87. Elbrönd, J.: Calculation of optimum size for ore bunkers (Berechnung der günstigsten Größe von Erzbunkern) (German). Glückauf 97 (1961), pp. 79 - 81.

88. Ellis, S. P. J.: Design of silos with aeration-assisted discharge. Symposium "Practical Bunker Design", IMechE, London, 25th February 1969, paper 18.

89. Engh, T. A.: A study of flow and blending in a homogenizer and storage silo system for the raw mix input in cement production. Chem. Engng. Lab., The Technical University of Norway, Trondheim 1966, pp. 1 - 15, 33 - 51, 115 - 119.

90. Engh, T. A.: Effect of injected air on the rate of flow of solids. ASME-Publ., No. 68-MH-18.

91. Ersov, L. V.; Lisin, A. A.: Methods to improve operations of bunker plants (Russian). Teploenergetika 11 (1964), pp. 44 - 47.

92. Ertle, H. J.: The bunker-discharge conveyor - a new and universal equipment for bulk masses (Der Bunkerabzugsförderer - ein neuartiges und universell einsetzbares Zwischenfördermittel für Massengüter) (German). Hebe- und Fördertechnik (1969), No. 9, pp. 41 - 45.

93. Fincham, L. W.: Material flow problems in bunkers and hopper feeds. Symposium "Practical Bunker Design", IMechE, London, 25th February 1969, paper 13.

94. Fisher, J. J.: Solid-solid blending. Chem. Engng. (1960), pp. 107-128.

95. Forward, P. J.: The application of electrical weighing to powder handling. Powder handling and storage, paper 12, Institute of Materials Handling, Maidenhead, Berksh., 1969, Ed.: A. S. Goldberg.

96. Foßhag, W.: Fluidizing systems in storage silos (Pneumatische Auflockerungssysteme in Vorratssilos) (German). Aufber. Techn. 6 (165), pp. 73 - 78.

97. Franke, E. W.: Discharge of bulk materials from bins by vibratory conveyor plants (Bunkerabzug von Schüttgütern mit Schwingförderanlagen) (German). Aufber. Techn. 10 (1969), pp. 579-583.

98. Franklin, F. C.; Johanson, L. N.: Flow of granular material through a circular orifice. Chem. Engng. Sci. 4 (1955), pp. 119 - 129.

99. Funk, K.: Effect of inserts on the flow of crushed limestone in a shaft kiln (Die Beeinflussung des Fließens von Kalksteinsplitt in einem Schachtofen durch Einbauten über dem Austrag)(German). III. CHISA Congress, Sept. 1969, Mariánské Lázně, CSSR, paper B 2.12.

100. Gardner, G. C.: The "best" hopper profile for cohesive material. Chem. Engng. Sci. 18 (1963), pp. 35 - 39.

101. Gardner, G. C.: The axisymmetric "best" hopper, its relevance to the plane "best" hopper and "best" hoppers integrated with bins. Chem. Engng. Sci. 19 (1964), pp. 283 - 288.

102. Gardner, G. C.: The region of flow when discharging granular materials from bin-hopper systems. Chem. Engng. Sci. 21 (1966), pp. 261 - 273.

103. Garve, T. W.: Segregation in bins. J. Am. Ceramic Soc. 8 (1928), pp. 666 - 670.

104. Giunta, J. S.: Flow patterns of granular materials in flat bottom bins. ASME-Publ., No. 68-MH-1.

105. Glahe, E.: Tests on bunkers for powdered coal (Versuche an Kohlestaubbunkern) (German). Energie 6 (1954), No. 3, p. 85.

106. Gleed, D. B.; Leesment, H. H.: Pebble bed reactor study flow visualization experiment. Atomic Energy Australia 8 (1965), No. 14, pp. 14 - 16.

107. Gorbis, Z. R.; Elkin, G. I.; Slyusarenko, N. N.: Flow rate measurement of granular materials. III. CHISA Congress, Sept. 1969, Mariánské Lázně, CSSR, paper B 5.7.

108. Grapengiesser, J. C.: A new type of aerating unit for aerated silo bottoms (Eine neuartige Belüftungseinheit für belüftete Silos) (German). Zement-Kalk-Gips 22 (1969), pp. 218 - 221.

109. Grothe, H.: Encyclopaedia of mining (Lexikon des Bergbaus) (German). Stuttgart: Deutsche Verlagsanstalt 1962, pp. 205 - 209.

110. Grünwald, F.: The bin discharge by means of discharging chutes with electromagnetic drive (Bunkerentleerung durch Abzugsrinnen mit elektromagnetischem Antrieb) (German). Aufber. Techn. 6 (1965), pp. 104 - 108.

111. Gumz, W.: Design of bunker outlets (Gestaltung von Bunkerausläufen) (German). Brennstoff-Wärme-Kraft 10 (1958), p. 122.

112. Guthrie, J. J.; Morris, J. M.: Electromechanical feeding and proportioning of bulk solids. ASME-IEEE Material Handling Engineering Conference, Oct. 18 - 20th 1965, Pittsburgh, Pa.

113. Hager, M.: Investigations on the feeding and recovery of bulk materials from bunkers with belt conveyors. (Untersuchungen zur Befüllung und zum Gutaustrag von Bunkern mit Abzugsbändern) (German). Dr. Thesis, Technical University Hannover, Germany, 1965.

114. Hampel, H. R.: Investigations on some factors influencing the behaviour of granular materials discharging out of a bunker (Experimentelle Untersuchungen einiger Einflußgrößen beim Ausfließen von Schüttgütern aus Bunkern) (German). Bergbauwissenschaften 5 (1958), pp. 313 - 327.

115. Handley, M. F.; Perry, M. G.: Measurements of stresses in flowing granular materials. Rheol. Acta 4 (1965), pp. 225 - 235.

116. Handley, M. F.; Perry, M. G.: Stresses in granular materials flowing in converging hopper sections. Powder Technol. 1 (1967/68), pp. 245 - 251.

117. Henny, G. E. J.: Bulk grain stores with particular reference to the design of vertical silos. Trans. South-African Inst. Civil Engrs. 7 (1957), pp. 225 - 242.

118. Herne, H.: Theory of solids handling. National Coal Board, London, 1964.

119. Heywood, H.: Calculation of the specific surface of a powder. Proc. Inst. Mech. Engrs. 125 (1933), pp. 383 - 459.

120. Hickerson, W. L.: Vibrating conveyor and feeder systems. ASME Vibrations Conference, 3rd Jan. 1967, Boston, Mass.

121. Hirst, A. A.: Bins for storage and proportioning of raw coal (Dosier- und Stapelbunker für Rohkohle) (German). 2nd Congress on Coal Preparation, Sept. 1954, Essen, Germany, paper A VI 2.

122. Hoffmann, E.: Discharge from slot bunkers (Schlitzbunkerausträge) (German). Bergbauwissenschaften 13 (1966), pp. 386 - 387.

123. Holmes, C. W. H.: The sampling of coal. Coll. Engng. (1934), pp. 40 - 42, 188 - 192.

124. Hoormann, W.; Hager, M.: Bin discharging devices. Stress, requirements, dimensioning, types (Bunkeraustragsorgane, Beanspruchung, Anforderungen, Dimensionierung, Bauarten) (German). Aufber. Techn. 8 (1967), pp. 521 - 537.

125. Horn, H.: Supralen RCH 1000 for lining bunkers and chutes (Supralen RCH 1000 für Bunker- und Rutschenauskleidung) (German). Hebe- und Fördertechnik (1970), No. 9, pp. 117 - 118.

126. Ivanov, I. G.: Investigation on the arch formation in bins and hoppers (Russian). Gosgortechizdat (1961), p. 98.

127. Janelid, I.; Kvapil, R.: Sublevel caving. Int. J. Rock Mech. Min. Sci. 3 (1966), pp. 129 - 153.

128. Janssen, H. A.: Experiments on grain pressures in silos (Versuche über Getreidedrücke in Silozellen) (German). VDI-Z. 39 (1895), pp. 1045 - 1049.

129. Jenike, A. W.; Elsey, P. J.; Woolley, R. H.: Flow properties of bulk solids. Univ. Utah, Engng. Exp. Station, Bull. No. 95, 1959.

130. Jenike, A. W.: Gravity flow of bulk solids. ASME-Publ., No. 60-WA-81.

131. Jenike, A. W.: A preliminary study of segregation. Univ. Utah, Engng. Exp. Station, Bull. No. 24, 1960.

132. Jenike, A. W.: Gravity flow of solids. Univ. Utah, Engng. Exp. Station, Bull. No. 108, 1961.

133. Jenike, A. W.: Storage and flow of solids. Univ. Utah, Engng. Exp. Station, Bull. No. 123, 1964.

134. Jenike, A. W.: Quantitative design of mass flow bins. Powder Technol. 1 (1967/68), pp. 237 - 244.

135. Jenike, A. W.; Johanson, J. R.: Bin loads. J. Structural Div. ASCE 94 (1968), ST4, paper 5916, pp. 1011 - 1041.

136. Jenike, A. W.; Johanson, J. R.: Solids flow in bins and moving beds. Chem. Engng. Progress 66 (1970), No. 6, pp. 31 - 34.

137. Johanson, J. R.; Colijn, H.: New design criteria for hoppers and bins. Iron and Steel Engr. (1964), Oct., pp. 85 - 104.

138. Johanson, J. R.: The use of flow-corrective inserts in bins. J. Engng. Ind. (1965), No. 65-MH-1.

139. Johanson, J. R.: New design concepts for coal bins and hoppers. Coal Age (1966), Jan., pp. 104 - 108.

140. Johanson, J. R.; Kleysteuber, W. K.: Flow corrective inserts in bins. Chem. Engng. Progress 62 (1966), No. 11, pp. 79 - 83.

141. Johanson, J. R.: Inserts to improve the flow of bulk materials in bins (Einbauten für Bunker und Silos zur Verbesserung der Schüttgut-bewegung) (German). Bergbauwissenschaften 14 (1967), pp. 85 - 90.

142. Johanson, J. R.: Effect of initial pressures on flowability of bins. ASME-Publ., No. 68-MH-9.

143. Johanson, J. R.: In-bin blending. Chem. Engng. Progress 66 (1970), No. 6, pp. 50 - 55.

144. John, P.: Storage, proportioning, and blending bins for raw coal (Stapel-, Dosier- und Mischbunker für Rohkohle). Aufber. Techn. 6 (1965), pp. 84 - 89.

145. Kalenderyan, V. A.; Titar, S. S.; Muratov, K. P.: Influence of vibration on the gravitational flow of a bed of granular materials. III. CHISA Congress, Sept. 1969, Mariánské Lázně, CSSR, paper B 3.5.

146. Karman, P. C.: Flow of gases through porous media. London 1956.

147. Kelka, W.: Bunker with several discharge screws for difficult flowing bulk products (Bunker mit mehreren Austragsschnecken für schwerfließende Schüttgüter). (German). Hebe- und Fördertechnik (1968), No. 10, p. 92.

148. Kelka, W.: Bins with movable feeder worms for the promotion of flow (Bunker mit fahrbaren Austragsschnecken für schwerfließende Schüttgüter). (German). Fördern und Heben 18 (1968), pp. 805 - 807.

149. Keneman, F. E., (Russian) Izv. Akad. Nauk SSR, Otd. Tekhn. Nauk Mekha i Mashinostr. (1960), No. 2, pp. 70 - 77.

150. Keneman, F. E., (Russian), Inzh. Fiz. Zh. 3 (1960), No. 4, pp. 18 - 22.

151. Khodak, L. Z.; Borisov, Y. I.: Velocity and pressure distribution of a granular material in a model shaft kiln. III. CHISA Congress, Sept. 1969, Mariánské Lázně, CSSR, paper B 2.11.

152. Kirwan, J. O.; Demler, L. E.: Continuous weighing and feeding. Instr. and Automat. 28 (1955), p. 98.

153. Klein, H.: Fundamental principles of pneumatic homogenization (Gesetzmäßigkeiten der pneumatischen Homogenisierung) (German). Zement-Kalk-Gips 15 (1962), pp. 399 - 402.

154. Kneschke, G.: Reducing arch formation when storing very fine grained bulk materials (Verringerung der Brückenbildung bei gebunkerten feinstkörnigen Schüttgütern) (German). Aufber. Techn. 9 (1968), pp. 603 - 608.

155. Koenen, M.: Calculation of wall and bottom pressures in silos (Berechnung des Seiten- und Bodendrucks in Silos) (German). Zentralblatt der Bauverwaltung 16 (1896), pp. 446 - 449.

156. Korn, A. H.: Practical pointers in design of storage bins. Chem. Engng. 53 (1946), Sept., pp. 100 - 101.

157. Koster, K. H.: Investigations on the flow behaviour of stowing material in bunkers fitted with cascade chutes (Experimentelle Untersuchungen der Vorgänge in Falltreppen beim Fördern und Bunkern von Versatzmaterial) (German). Bergbauwissenschaften 8 (1961), pp. 125 - 139.

158. Koster, K. H.: Open air storage of bulk material (Speicherung von Schüttgut im Freilager) (German). Bergbauwissenschaften 13 (1966), pp. 378 - 382.

159. Kott, A.; Kramer, F. B.; Influence of wall material on bulk solids flow. Chem. Engng. Progress 62 (1966), No. 11, pp. 73 - 78.

160. Kühn, H.: A new bunker for non-flowing materials (Ein neuer Bunker für schlecht fließende Schüttgüter) (German). Bergbauwissenschaften 14 (1967), pp. 91 - 94.

161. Kühn, H. : Homogenizing of bulk products by means of stock piles with rotary-plow feeders (Das Homogenisieren von Schüttgütern mit Hilfe von Zwischenlagern mit Schlitzbunker-Entleerungswagen) (German). Hebe- und Fördertechnik 13 (1967), pp. 491 - 496.

162. Kusnecov, R. F. : Arch formation of coarse materials in shaft kilns (Russian). Symposium "Storage of Bulk Materials", Odessa 1967.

163. Kvapil, R. : Movement of granular materials in bins and bunkers (Schüttgutbewegungen in Bunkern) (German). Berlin: VEB Verlag Technik 1959.

164. Kvapil, R. : Problems of designing bunkers for free and non-free flowing bulk materials (Problematik der Konstruktion von Bunkern) (German). Bergbautechnik 10 (1960), pp. 466 - 471.

165. Kvapil, R. : Construction of bunkers and chutes for coarse materials (Konstruktionsprobleme bei Entladetrichtern, Bunkern und Rollen für grobstückiges Gut) (German). Bergakademie 13 (1961), pp. 451 - 459.

166. Kvapil, R. : Problems of the gravity flow of bulk materials (Probleme des Gravitationsflusses von Schüttgütern) (German). Aufber. Techn. 5 (1964), pp. 139 - 144, 183 - 189.

167. Kvapil, R. : Specific problems of the gravity flow of bulk materials (Besondere Probleme der Gravitationsbewegung von Schüttgütern) (German). Aufber. Techn. 5 (1964), pp. 544 - 549, 642 - 654.

168. Kvapil, R. ; Tanaka, T. : Bins and bunkers with shield inserts (Bunker und Silos mit Schild) (German). Aufber. Techn. 6 (1965), pp. 45 - 49.

169. Kvapil, R. : Gravity flow of granular materials in hoppers and bins. Int. J. Rock Mech. Min. Sci. 2 (1965), pp. 25 - 41, 277 - 304.

170. Kvapil, R. : Gravity flow of bulk materials in bunkers (Bunkerung aus der Sicht des Gravitationsflusses) (German). Bergbauwissenschaften 13 (1966), pp. 327 - 329.

171. Kvapil, R. ; Taubmann, H. : Flow and extraction of solids from bins. ASME-Publ., No. 68-MH-32.

172. Laforge, R. M. ; Boruff, B. K. : Profiling flow of particles through hopper openings. Ind. Engng. Chem. 56 (1964), No. 2, pp. 42 - 46.

173. Lambert, F. W.: The theory and practical design of bunkers. London: British Constructional Steelwork Ass. Ltd., Publ.-No. 32, 1968.

174. Lange, P. A.; Künnecke, M.: An evaluation of pneumatical homogenizers in the cement industry (Über die Bewertung von pneumatischen Homogenisieranlagen in der Zementindustrie) (German). Tonind.-Ztg. 88 (1964), pp. 424 - 426.

175. Lange-Eichholz, K. H.; Wasow, H.: Comments on the design of bunkers (Bemerkungen zur Konstruktion von Bunkeranlagen) (German) Erzmetall 7 (1954), pp. 459 - 461.

176. Langmaid, R. N.; Rose, H. E.: Arch formation in a non-cohesive granular material. J. Inst. Fuel 30 (1957), No. 195, pp. 166 - 172.

177. Lebedinski, V. G.: The resistance of inserts of various geometric shape in a bulk material stream (Widerstand von Körpern verschiedener geometrischer Form, die im Strom des Schüttgutes angebracht sind) (Russian). Symposium "Storage of Bulk Materials", Odessa 1967.

178. Lee, C. A.: New ideas about hoppers. Chem. Engng. 59 (1952), April, pp. 153, 173.

179. Lee, C. A.: Design of hoppers for use. Chem. Engng. 60 (1953), May, pp. 194 - 195, 200.

180. Lee, C. A.: Hoppers by calculation? Chem. Engng. 61 (1954), Dec., p. 181.

181. Lee, C. A.: Solids feeders. Chem. Engng. 71 (1964), Febr., pp. 119 - 134.

182. Lee, Y.: Flow of coal in bunkers. Combustion 31 (1960), No. 7, pp. 20 - 27.

183. Lee, Y.: Flow of coal in bunkers (Kohlefluß in Bunkern) (German). Arch. Energiewirtsch. (1960), pp. 417 - 434.

184. Legget, R. F.: Clogging of bitumious coal in bunkers. Trans. ASME (1947), July, pp. 525 - 533.

185. Lemke, K.; Kröner, H.; Rösner, W.: Breakage of coal during dropping and bunkering (Die Zertrümmerung von Kohle beim Bunkern und Stürzen) (German). Glückauf 102 (1966), pp. 1129 - 1139.

186. Léon, K. W.: Bunker design and discharge (Formgebung und Entleerung von Bunkern) (German). Brennstoff-Wärme-Kraft 9 (1957), pp. 424 - 427.

187. Leva, M.; Grummer, M.; Weintraub, G.; Pollchik, M.: Fluidization of an anthracite coal. Ind. Engng. Chem. 41 (1949), p. 1206.

188. Lightfoot, E.; Michael, D.: Prismatic coal bunkers in structural steelwork. Structural Engr. 44 (1966), Febr.

189. Manecke, H.: Tests carried out with a Fuller-homogenizer (Versuche mit einem Fuller-Homogenisiersilo) (German). Zement-Kalk-Gips 7 (1954), p. 464.

190. Markov, K. D.; Girzel, A. M.: Prediction of wear of coal bunker walls (Russian). Sachtnstroit (1969), No. 3, pp. 18 - 21.

191. Martens, P.: Silo loads for powdered materials and by aeration (Silolasten aus staubförmigen Schüttgütern und aus Luftzufuhr) (German). Dr. Thesis, Technical Univ. Braunschweig, Germany, 1969.

192. Martens, P.: Behaviour of powdered materials with and without aeration (Verhalten von staubförmigen Schüttgütern mit und ohne Luftzufuhr) (German). Bautechnik (1970), No. 3, pp. 77 - 80.

193. Martin, J. B.; Richards, J. C.: The determination of the dynamic zone within a free-flowing granular mass. J. Sci. Technol. 11 (1965), No. 1, pp. 31 - 34.

194. Matthée, H.: Mathematical determination of discharge quantities from bunkers and the method by Jenike (Zur formelmäßigen Erfassung von Austragsmengen aus Bunkern - das Verfahren von Jenike) (German). Bergbauwissenschaften 13 (1966), pp. 333 - 341.

195. Matthée, H.: Segregation phenomena relating to bunkering of bulk materials. Powder Technol. 1 (1967/68), pp. 265 - 271.

196. Matthée, H.: Findings on the segregation of bead fillings in model bunkers and their applicability to industrial conditions using the similarity theorem (Erkenntnisse über die Entmischung von Kugelschüttungen in Bunkern und ihre Übertragbarkeit auf Industrieanlagen mit Hilfe der Ähnlichkeitslehre) (German). Bergbauwissenschaften 15 (1968), pp. 443 - 457.

197. Mayer, F. W.: Segregation as a dressing process (Der Entmischungsvorgang als physikalischer Aufbereitungsprozess) (German). Bergb.-Archiv 11/12 (1950), pp. 82 - 94.

198. Mazza, M.: Equipments for bunker discharge (Bunkerabzugsvorrichtungen) (German). Hebe- und Fördertechnik 7 (1961), No. 5, pp. 22 - 26.

199. v. Mildenstein, M.: Plastic linings for silos and other containers (Kunststoffbeschichtungen für Silos und andere Großbehälter) (German). Aufber. Techn. 6 (1965), pp. 96 - 100.

200. Mitchell, D. R.: Segregation in the handling of coal. Trans. AIME 130 (1938), pp. 107 - 142.

201. Möller, H.: Discharge from silos and loading-in bulk of pulverized and fine-grained material (Siloabzug und lose Verladung von feinkörnigem Schüttgut) (German). Aufber. Techn. 6 (1965), pp. 65 - 72.

202. Mößner, A. P.: Proposals for regular discharge of storage bins (Vorschläge für die gleichmäßige Entleerung von Vorratsbehältern) (German). Glückauf 71 (1935), pp. 1130 - 1131.

203. Müller, H.: Methods of bin level measurement and their application (Möglichkeiten und Grenzen der Verfahren zur Bunkerstandsmessung) (German). Glückauf 98 (1962), pp. 1357 - 1369.

204. Musschoot, A.: Vibratory feeders and conveyors: designs, new concepts and applications. ASME Annual Meeting, Nov. 1963, Philadelphia, Penn.

205. Myers, J. L.: New method of applying vibrating hoppers. ASME-Publ., No. 69-MH-16.

206. Myers, J. L.: Vibrating hoppers. Mech. Engng. (1970), March, pp. 27 - 31.

207. Nakajima, H.; Gotoh, K.; Tanaka, T.: Instantaneous flow meter for granular materials and its design. III. CHISA Congress, Sept. 1969, Mariánské Lázně. CSSR, paper B 5.8.

208. Nakamura, Y.: More efficient powder flow by means of vibromotor (Japanese). J. Res. Assoc. Powder Tech. 4 (1967), No. 1, pp. 671 - 676.

209. Nakamura, Y.: Prevention of arch formation in bins and hoppers. Chem. Engng. Tokyo, 13 (1969), No. 4.

210. Nanninga, N.: Does the conventional method for calculating bunker loads give sufficient results? (Dutch). De Ingenieur 68 (1956), Nov.

211. Newton, R. H.; Dunham, G. S.; Simpson, T. P.: The TCC catalytic cracking process for motor gasoline production. Trans. Am. Inst. Chem. Engrs. 16 (1945), pp. 215 - 232.

212. Novosad, J.: Studies on granular materials. Collection Czechoslov. Chem. Comm. 29 (1964), pp. 2681 - 2709.

213. Novosad, J.: Residence time distribution of granular materials flowing through cylindrical vessels. THTR Symposium "Problems of the Pebble Bed and Granular Materials", Jülich, Germany, March 1968, EURATOM-Publ., No. EUR 4190 d, e.

214. O'Callaghan, J. R.: Internal flow of moving beds of granular material. J. Agricult. Engng. Res. (1959), No. 4, pp. 200 - 217.

215. Olive, T. R.: Solids feeders. Chem. Engng. 59 (1952), No. 11, pp. 163 - 178.

216. Oszter, Z. F.: Comments on specific feeder applications. Canad. Min. Met. Bull. 59 (1966), March., pp. 363 - 382.

217. Oszter, Z. F.: Improving material flow in plug-flow type fine ore bins - case studies. CIM Bull. (1970), March, pp. 333 - 338.

218. Otte, W.: Movement of crowds of people through narrow passes (Strömung von Menschenmengen durch Engpässe) (German). VDI-Z. 95 (1953), p. 240.

219. Pajer, G.; Kurth, F.: Continuous conveyors (Stetigförderer) (German). Berlin: VEB Verlag Technik 1967, p. 230.

220. Pariseau, W. G.: Gravity flows of ideally plastic materials through slots. Trans ASME (1969), May, pp. 414 - 422.

221. Pariseau, W. G.: Discontinuous velocity fields in gravity flows of granular materials through slots. III. CHISA Congress, Sept. 1969, Mariánské Lázně, CSSR, paper B 3.8.

222. Pawlowski, J.: The statistics of homogenizing processes (Zur Statistik der Homogenisierprozesse) (German). Chem. Ing. Techn. 36 (1964), pp. 1089 - 1098.

223. Peacock, H. M.: The design or adaption of storage bunkers to prevent size segregation of solids. J. Inst. Fuel 11 (1938), pp. 230 - 239.

224. Peschl, I. A. S. Z.: Theory of the formation of arches in bins. ASME-Publ., No. 68-MH-4.

225. Peschl, I. A. S. Z.: Arch formation in bins and bunkers (Brücken- bzw. Gewölbebildung bei der Schüttgutspeicherung in Bunkern) (German). Bergbauwissenschaften 17 (1970), pp. 89 - 95, 127 - 134, 263 - 265.

226. Peters, K.; Schmidt, A.: Investigations on fluidized beds (Untersuchungen über das Wirbelschichtverfahren) (German). Öst. Chem. Z. 54 (1953), pp. 253 - 258.

227. Pieper, K.; Wenzel, F.: Comments on DIN 1055, page 6 - design loads for buildings; loads in silo bins (Erläuterungen zu DIN 1055, Blatt 6 - Lastannahmen für Bauten; Druckverhältnisse in Silozellen) (German). Beton und Stahlbetonbau (1963), pp. 6 - 11.

228. Pieper, K.; Wenzel, F.: Pressure conditions in silos (Druckverhältnisse in Silozellen) (German). Berlin: W. Ernst & Sohn, 1964.

229. Pieper, K.: Investigation of silo loads in measuring models. ASME-Publ., No. 68-MH-30.

230. Pieper, K.; Wagner, K.: The influence of different types of discharging on the side pressures in silo compartments (Der Einfluß verschiedener Auslaufarten auf die Seitendrücke in Silozellen) (German). Aufber. Techn. 10 (1969), pp. 542 - 546.

231. Pirie, J. M.: The handling of solids. London: Inst. Chem. Engrs. 1962.

232. Ponomarenko, N. I.: The influence of shape and lining of bunkers on the degradation during discharge (Russian). Symposium "Storage of Bulk Materials", Odessa 1967.

233. Poulton, B. A.: Hopper design for the bin activator. Symposium "Practical Bunker Design", IMechE, London, 25th February 1969, paper 16.

234. Quin, P. : Power station coal bunker outlets that have been re-designed to improve flow. Symposium "Practical Bunker Design", IMechE, London, 25th February 1969, paper 7.

235. Rachner, H.-G. : Vibrating conveyor plants for bin feeding and discharging (Schwingförderanlagen zum Beschicken und Entleeren von Bunkern) (German). Aufber. Techn. 6 (1965), pp. 109 - 114.

236. Radewald, H. : State-of-the-art of blending of raw material in a cement production plant (Entwicklung und Stand der Homogenisierung von Rohmaterial und Rohmehl in einem Zementwerk) (German). Zement-Kalk-Gips 22 (1969), p. 184.

237. Radewald, H. : Problems of the homogenization of raw mix in Harburg (Probleme der Homogenisierung von Rohmehl in Harburg) (German). Zement-Kalk-Gips 22 (1969), pp. 371 - 374.

238. Raymond, J. L. : Stainless steel coal ducts stop trouble in maintenance and operation. Power Engng. 59 (1955), May, p. 95.

239. Rehm, H. : Electromagnetic vibrating conveyor pans (Elektromagnetische Vibrationsförderrinnen) (German). Glückauf 106 (1970), pp. 328 - 330.

240. Reimbert, M. ; Reimbert, A. : Bunkers, theory and practical applications (Silos, traité théorique et pratique) (French). Paris: Edition Eyrolles 1959. Silos (German). Wiesbaden: Bauverlag 1961.

241. Reisner, W. : Experiments on flow and pressure behaviour of stored materials in relation to the shape of bunkers (Beitrag zur Untersuchung der Fließ- und Druckverhältnisse von gebunkerten Schüttgütern in Abhängigkeit von der Bunkerform) (German). Bergbauwissenschaften 8 (1961), pp. 175 - 189.

242. Reisner, W. : Additional equipment and special bunker shapes to improve the movement of bulk materials (Zusatzeinrichtungen und Sonderformen von Bunkern für bessere Schüttgutbewegungen) (German). Bergbauwissenschaften 13 (1966), pp. 371 - 377.

243. Reisner, W. : The behaviour of granular materials in flow out of hoppers. Powder Technol. 1 (1967/68), pp. 257 - 264.

244. Reisner, W. : Parameters to characterize flow behaviour of spherical and non-spherical materials. ASME-Publ., No. 68-MH-40.

245. Reisner, W.: Possibilities for improving the movement of bulk materials in bunkers. III. CHISA Congress, Sept. 1969, Mariánské Lázně, CSSR, paper B 2.8.

246. Reynolds, W. J.; Carter, R. S.: Design of bunkers for difficultly flowing fuels. Combustion (1962), Sept., pp. 32 - 36.

247. Richards, J. C.: Granular flow through bunkers. Brit. Coal Util. Res. Ass. 19 (1955), No. 4, pp. 145 - 157.

248. Richards, J. C.: The importance of some apparently minor features in bunker design. Brit. Coal Util. Res. Ass. 27 (1963), No. 2, pp. 33 - 48.

249. Richards, J. C.: The storage and recovery of particulate solids. London: Inst. Chem. Engrs. 1966.

250. Richards, J. C.: Handling bulk solids - a framework of concepts. Particulate Matter 1 (1970), No. 2, pp. 15 - 20.

251. Richmond, O.: Gravity hopper design. ASME-Publ., No. 62-WA-42.

252. Richmond, O.: Bunker design for gravity flow (Gestaltung von Bunkern mit Schwerkraftdurchsatz). Energie 15 (1963), pp. 214 - 216.

253. Rischel, H.: Application of slot bunker with horizontal shelf and rotary-plow feeder (Der Schlitzbunker mit horizontalem Austragstisch und Entleerungswagen und seine Anwendungsmöglichkeiten)(German). Zement-Kalk-Gips 16 (1963), pp. 267 - 274.

254. Rösner, W.; Lemke, K.: Measures for the careful handling of coal during bunkering and reloading (Maßnahmen zum Schonen von Steinkohle beim Bunkern und beim Umschlagen)(German). Glückauf 105 (1969), pp. 147 - 155.

255. Röttger, H.: Transloading bulk material on continuous conveyors (Schüttgutübergabe auf Stetigförderer) (German). Düsseldorf: VDI-Verlag 1964.

256. Rose, H. E.; Tanaka, T.: Rate of discharge of granular materials from bins and hoppers. The Engineer (1959), pp. 465 - 469.

257. Runo, W. R.: How orifice shape affects material flow. Modern Mat. Handl. (1956), Febr., pp. 90 - 91.

258. Ruth, M.: Practical experiences with the blending of raw material (Betriebserfahrungen mit der selbständigen Rohmaterialmischung) (German). Zement-Kalk-Gips 22 (1969), p. 185.

259. Sachmeister, L. G.; Jaroŝewski, B. I.: Calculation of bunker size for intermediate storage of coal underground by means of the queuing theory (Russian). Ugol 42 (1967), No. 8, pp. 66 - 68.

260. Salzer, G.: Haulage equipment for bulk materials (Schüttgutförderer) (German). Mainz: Krausskopf Verlag 1968, p. 200.

261. Sanders, P. M.: Handling paper coating materials. Mech. Handl. (1962), Sept.

262. Sandstrom, C. O.: Avoiding clogged bin hoppers. Chem. Met. Engng. 47 (1940), No. 1, pp. 22 - 23.

263. Schieck, E.: Flow of bulk materials from asymmetric bunkers (Beitrag zum Ausflußverhalten von Schüttgütern aus unsymmetrischen Bunkern) (German). Bergakademie 18 (1966), pp. 153 - 156.

264. Schmidt-Holthausen, H.-J.: Experiences in the storage of industrial dusts (Erfahrungen beim Bunkern von Nutzstäuben)(German). Staub 27 (1967), pp. 91 - 95.

265. Schrämli, W.: On the measurement of the flow properties of cement. Powder Technol. 1 (1967/68), pp. 221 - 227.

266. Schultheis, W.: Vibrating conveyors in the preparation and process technology (Schwingförderer in der Aufbereitung und Verfahrenstechnik) (German). Aufber. Techn. 10 (1969), pp. 441 - 445.

267. Schwedes, J.: Flow of bulk materials in bunkers (Fließverhalten von Schüttgütern in Bunkern) (German). Weinheim: Verlag Chemie 1968.

268. Schwedes, J.: Dimensioning of bins (Dimensionierung von Bunkern) (German). Aufber. Techn. 10 (1969), pp. 535 - 541.

269. Seiler, K.: Practical experiences with devices for level measurement in bunkers, weighing and dosing of bulk materials and controlling of pulp density in lime plants (Praktische Erfahrungen mit Meßgeräten zur Messung von Bunkerständen, zum Wiegen und Dosieren von Schüttgütern und zur Trübemengenmessung in der Kalkindustrie) (German). Erzmetall 20 (1967), pp. 569 - 574.

270. Shelby, A. T.: The application of mass flow theory to a ground phosphate problem. 72nd Annual General Meeting of the Canadian Institute of Mining, Toronto, April 1970.

271. Sherman, R. A.; Kaiser, E. R.: Segregation of coal in an industrial steam plant bunker. Combustion (1937), Dec., pp. 25 - 28.

272. Shimon, D. F.: Bin design. Feed Production Handbook. Feed Production School Inc. 1961, pp. 271 - 275.

273. Shinohara, K.: Designing hoppers for cohesive material. Min. Proc. (1967), April, pp. 26 - 30.

274. Shinohara, K.; Suzuki, A.; Tanaka, T.: Gravity and vibration effects on flow of cohesive materials from hoppers. ASME-Publ., No. 68-MH-33.

275. Sinden, A. D.: Getting difficult materials out of bins. ASME-Publ., No. 61-BSH-8.

276. Small, R. L.; Colijn, H.: Belt scale weighing can be accurate. Instrum. Soc. Amer., Preprint 11.3 - 1 - 64.

277. Sommer, H.: Bins for dewatering fine materials (Bunkerung für Entwässerung feinkörniger Güter) (German). Aufber. Techn. 6 (1965), pp. 101 - 103.

278. Soth, W.: Experiences with bin linings (Erfahrungen mit Bunkerauskleidungen) (German). Aufber. Techn. 10 (1969), pp. 595-598.

279. Spiwakowski, A. O.; Djatschkow, W. K.: Haulage equipment - transport and handling systems for bulk materials (Förderanlagen - Transport- und Verladeeinrichtungen für Massengüter) (German). Braunschweig: Friedr. Vieweg & Sohn Verlag; Berlin: VEB-Verlag Technik 1959.

280. Stepanoff, A. J.: Gravity flow of bulk solids. Hopper flow charakteristics. ASME-Publ., No. 68-MH-34.

281. Stepanoff, A. J.: Gravity flow of bulk solids and transport of solids in suspensions. New York: John Wiley & Son Inc. 1969.

282. Straußfeld, R.; Velhagen, F.: Planning, construction, and operational experiences with a homogenizing bin (Planung, Aufbau und Betriebserfahrungen mit einem Vergleichmäßigungsbunker) (German). Aufber. Techn. 10 (1969), pp. 569 - 578.

283. Stoyan, D. : A mathematical model of variations of excavator and overburden-spreader output and its application to the design of bunkers in belt operation (Ein mathematisches Modell der Leistungsschwankungen bei Baggern und Absetzern und seine Anwendung bei der Dimensionierung von Bunkern in Bandbetrieben) (German). Bergbautechnik 17 (1967), pp. 403 - 406.

284. Stoyan, D. : On designing bins and bulk stores for stationary feed and discharge (Ein Beitrag zum Problem der Dimensionierung von Bunkern und Massenspeichern bei stationärer Zufuhr und Entnahme) (German). Bergbautechnik 18 (1968), pp. 582 - 586.

285. Stumpf, K. : Measurement of bin contents (Die Inhaltsmessung von Bunkern) (German). Aufber. Techn. 6 (1965), pp. 90 - 95.

286. Swierkot, G. ; Taciuk, W. : Problems associated with the flow of bulk solids at the Iron Ore Comp. of Canada. 72. Annual General Meeting of the Canad. Inst. Min. , Toronto, April 1970.

287. Takahashi, K. : Experimental investigations on the velocity of efflux of granular mass. Sci. Pap. Inst. Phys. Chem. Res. Tokyo 26 (1935), pp. 11 - 20.

288. Tanaka, T. : What do you know about bin design? Rock Prod. (1961), Febr., pp. 115 - 126.

289. Tanaka, T. ; Kawai, S. : Hoppers can be used as mixers. Rock Prod. (1962), April, pp. 120 - 124.

290. Tarjahn, G. : On power effects of pelletizing (Der Einfluß von Bewegungskräften auf die Granulation) (German). Aufber. Techn. 7 (1966), pp. 28 - 32.

291. Taubmann, H. : Feeders and dischargers for bunkers with special aspects on electromagnetic vibrators (Austragsorgane und Austragshilfen für Bunker unter besonderer Berücksichtigung elektromagnetischer Vibratoren) (German). Chem. Ing. Techn. 28 (1956), pp. 250 - 257.

292. Taubmann, H. : Problems of storing bulk materials (Probleme der Speicherung von Schüttgütern) (German). Fördern und Heben 8 (1958), pp. 275 - 285.

293. Taubmann, H. : On the storage of bulk materials (Beitrag zur Speicherung von Schüttgut) (German). Aufber. Techn. 1 (1960), pp. 17-26.

294. Teichmann, E.: The discharge of bulk materials from bunkers (Austrag von Schüttgut aus Bunkern) (German). Zement-Kalk-Gips 16 (1963), pp. 259 - 266.

295. Theimer, O.: The segregation problem in grain silos (Das Problem der Entmischung von Getreide in Silozellen) (German). Deutsche Müller Z. 54 (1956), pp. 462 - 465.

296. Theimer, O.: A silo with cross-bottom hopper (Weserhütte Hameln) (Der Kreuzboden-Mehlsilo der Weserhütte Hameln) (German) Mühle 96 (1959), No. 29.

297. Theimer, O.: Flow promoting hopper construction (Ablauf fördernde Trichterkonstruktionen) (German). Aufber. Techn. 10 (1969), pp. 547 - 556.

298. Theimer, O.: Flow promoting devices and hopper constructions of silo bins (Ablauf fördernde Trichterkonstruktionen von Silozellen) (German). Powder Technol. 3 (1969/70), pp. 235 - 248.

299. Thirsk, A. C. S.: The bulk handling of flour and wheatfeed. Milling (1962), Aug.

300. Tiefenböck, H.: Level measurement by capacitance (Füllstandsmessung nach der kapazitiven Methode) (German). Elektrowelt (1965), No. B. 19.

301. Tomlin, N.: Spiral chute design for minimum degradation. Kings College Min. Soc. J. (1953), May.

302. Trenkler, H.: Shapes of bunkers for bulk materials (Formgebung von Bunkern für Schüttgüter, insbesondere für Kesselkohle) (German). Informations VGB 96 (1965), June, pp. 193 - 200.

303. Uetz, H.; Hakenjos, V.: Experiments on abrasion by slide-friction and slide-abrasion on plastics (Gleitreibungs- und Gleitverschleiß-versuche an Kunststoffen) (German). Kunststoffe 59 (1969), No. 3.

304. Valenti, F.: The automation of free flowing solids. Can. Min. Met. Bull. 45 (1961), p. 833.

305. Valentin, F. H. H.: The flow of bulk solids - what recent advances mean to the practical engineer. Powder handling and storage, paper 2, Institute of Materials Handling, Maidenhead, Berksh., 1969, Ed.: A. S. Goldberg

306. Vitunac, E. A.: Economic considerations in hopper design. Chem. Engng. Progress 62 (1966), No. 11, pp. 69 - 72.

307. Wakabayashi, T.: Observation of the stress pattern in powdered mass and its variation (Japanese). J. Res. Assoc. Powder Tech. 4 (1967), No. 1, pp. 657 - 670.

308. Walker, D. M.: An approximate theory for pressures and arching in hoppers. Chem. Engng. Sci. 21 (1966), pp. 975 - 997.

309. Walker, D. M.: Pressures and flow in coal bunkers. Powder Technol. 1 (1967/68), pp. 228 - 236.

310. Walter, L.: Automatic bulk material bunker features preselected flow rate (Automatischer Bunker für vorgewählten Fördergutstrom) (German). Fördern und Heben 14 (1964), p. 764.

311. Walvoord, O. W.: Choosing ore feeders for benefication plants. Min. Engng. (1955), Febr., pp. 131 - 134.

312. Weislehner, G.: Characteristic values for the assessment of pneumatic blending (Kenngrößen für die Beurteilung der pneumatischen Mischung) (German). Zement-Kalk-Gips 16 (1963), pp. 519 - 520.

313. Weislehner, G.: Considerations on the design of a mixing bed for pulverulent materials (Betrachtungen über die Auslegung eines Mischbettes für staubförmige Güter) (German). Zement-Kalk-Gips 18 (1965), pp. 549 - 552.

314. Weislehner, G.: The pneumatic mixing of raw mix in the cement industry (Die pneumatische Mischung von Rohmehl in der Zementindustrie) (German). Aufber. Techn. 10 (1969), pp. 561 - 568.

315. Wessmann, J. H.: Theory of screw conveyors. Feed Production Handbook. Feed Production School Inc. 1961, pp. 260 - 262.

316. Weydanz, W.: Time function of a blending process (Zeitlicher Ablauf eines Mischvorgangs) (German). Chem. Ing. Techn. 32 (1960), pp. 343 - 349.

317. Wheeler, E.: Bunker design. Symposium "Practical Bunker Design", IMechE, London, 25th February 1969, paper 17.

318. Whitaker, J. A.: The "Langwith" rapid loading chute, NCB patent No. 1099354. Symposium "Practical Bunker Design", IMechE, London, 25th February 1969, paper 9.

319. Wiegmann, D. : The pneumatic homogenization of pulverized materials (Die pneumatische Homogenisierung pulverförmiger Stoffe) (German). Aufber. Techn. 6 (1965), pp. 79 - 83.

320. Williams, J. C. : The segregation of powders and granular materials. Fuel Soc. J. 14 (1963), pp. 29 - 34.

321. Williams, J. C. : Cause and effects of segregation in powders. Chem. Process. Suppl. (1965), April.

322. Williams, J. C. ; Birks, A. H. : The preparation of powder specimens for shear cell testing. Rheol. Acta 4 (1965), pp. 170 - 180.

323. Williams, J. C. ; Shields, G. : The segregation of granules in a vibrated bed. Powder Technol. 1 (1967/68), pp. 134 - 142.

324. Williams, J. C. ; Birks, A. H. : The comparison of the failure measurements of powders with theory. Powder Technol. 1 (1967/68), pp. 199 - 206.

325. Wöhlbier, H. ; Reisner, W. : The storage of medium and fine grained bulk materials (Fragen der Bunkerung von mittel- und feinkörnigem Schüttgut) (German). Chem. Ing. Techn. 34 (1962), pp. 603 - 609.

326. Wöhlbier, H. ; Reisner, W. : Basic knowledge relating to the bunkering of bulk materials (Grundlegende Erkenntnisse bei der Bunkerung von Schüttgütern) (German). Fördern und Heben 13 (1963), pp. 406 - 416.

327. Wolf, E. F. ; v. Hohenleiten, H. L. : Experimental study of the flow of coal in chutes at Riverside Generating Station. Trans. ASME (1945), pp. 585 - 599.

328. Wright, H. : Bunker design for successful operations. Symposium "Practical Bunker Design", IMechE, London, 25th February 1969, paper 4.

329. Zaolarek, J. : Additional bunkers for powdered coal (Zwischenbunker für Kohlenstaub) (German). Brennstoff-Wärme-Kraft 9 (1957), pp. 427 - 429.

330. Zherdev, M. P. : Automation of the pneumatic loosening of coals in bunkers. Coke & Chem. USSR (Coal Tar R. L. Transl.) (1960), pp. 12 - 13.

331. Ziems, D.: Dimensioning of transport and storage units for transloading points. 5th Conference on Material Handling, Budapest, Oct. 1967, paper 59.

332. Ziller, A. H.; Löser, R.: Determination of pressures in bulk materials for finding the best hopper shape for a bin (Die Bestimmung der Drücke in Schüttgütern zum Festlegen der optimalen Ausflußform für Bunker) (German). Hebe- und Fördertechnik 12 (1966), No. 9, pp. 59 - 62.

333. Ziller, A. H.; Löser, R.: Mathematical example for the design of a hopper with hyperbolic walls (Rechenbeispiel zur Auslegung eines Bunkers mit hyperbolischen Wänden) (German). Hebe- und Fördertechnik 12 (1966), No. 12, pp. 63 - 66.

List of Manufacturers

In the following list only companies are mentioned which have contributed to this volume by making available information and data on their products. The list, therefore, is not intended to be complete.

Manufacturers not listed here are asked to add the names of the authors to their mailing list for future reference and evaluation in forthcoming new editions of this book.

1. AEG-Telefunken (E 55). D 6 Frankfurt Niederrad, Goldsteinstr. 238, Germany.
2. AEG-Werk Heiligenhaus, D 5628 Heiligenhaus, Germany.
3. Adolf Baumgarten Mühlen- und Speicherbau, D 4952 Hausberge a.d. Porta, Postfach 64, Germany.
4. B-I-F Industries Inc., P.O. Box 276, Providence 1, R.I., USA.
5. Bin Dicator Comp., 17190 Denver, Detroit, Michigan 48224, USA.
6. Brabender Technologie KG, D 41 Duisburg 1, Postfach 204 - 205, Germany.
7. Maschinenfabrik Karl Brieden & Co., D 463, Bochum, Postfach 50013, Germany.
8. Maschinenfabrik Buckau R. Wolf AG, D 4048 Grevenbroich, Lindenstr. 43, Germany.
9. Gebrüder Bühler AG, CH 9240 Uzwil, Switzerland.
10. Büttner-Schilde-Haas AG - Werk Grün -, D 6471 Lissberg/Oberhessen, Postfach, Germany.
11. Callow Engineering Ltd., Skelmersdale, Lancashire, Great Britain.
12. Carman Industries Inc., 1005 W. Riverside Drive, Jeffersonville, Indiana 47130, USA.

13. Claudius Peters AG, D 2 Hamburg 39, Kapstadtring 1, Germany.

14. Cleveland Vibrator Co., 2828 Clinton Ave., Cleveland, Ohio 44113, USA.

15. Clouth Gummiwerke AG, D 5 Köln 60, Niehler Str. 92 - 116, Germany.

16. Cowlishaw Walker & Co. Ltd., Biddulph, Stoke-on-Trent, Great Britain.

17. Crawley Engineering Developments Ltd., Llanelly, Carmarthenshire, Great Britain.

18. Davy and United Instruments Ltd., Darnall Works, Sheffield 9, Great Britain.

19. EMA-Dörries GmbH, D 519 Stolberg, Brockenweg 19, Germany.

20. Endress + Hauser GmbH + Co., D 7867 Maulburg, Germany.

21. Fairway Engineering Co. Ltd., Harcourt Street, Walkden, Worsley, Manchester M 28 5 GP, Great Britain.

22. Flottmann Werke GmbH, D 469 Herne, Postfach 240/260, Germany.

23. FMC Corporation, Syntron Division, Homer City, Pa. 15748, USA.

24. Förderanlagen und Maschinenbau GmbH, D 294 Wilhelmshaven, Banter Weg 11, Germany.

25. B. F. Gump Co., 1325 Cicero Ave., Chicago, Ill., USA.

26. Karl Händle & Söhne Maschinenfabrik, D 7130 Mühlacker, Postfach 209, Germany.

27. Hardinge Co., York, Pa., USA.

28. Hewitt Robbins Inc., Stamford, Conn. 06906, USA.

29. Hugo Herzbruch & Söhne, D 43 Essen-Bredeney, Alfredstr. 307, Germany.

30. Hosokawa Iron Works Ltd., Osaka, Japan.

31. R. Horsfield & Co., Leeds, Great Britain.

32. Industriebau Nord GmbH & Co. KG - Abt. Silotechnik -, D 2 Hamburg 70, Nebendahlstr. 11, Germany.

33. Krauss-Maffei AG, D 6 Frankfurt/M. 90, Postfach 900407, Germany.

34. Link Belt Co., 300 W. Pershing Rd., Chicago, Ill. 60609, USA.

35. Long Airdox Comp., Division of Marmon Group Inc., P. O. Box 331, Oak Hill, W. Va. 25901, USA.

36. Maschinenbau Louise GmbH, D 505 Porz-Westhoven, Industriestr. 32, Germany.

37. Louise Engineering Ltd., 53 Oldford Crescent, Acklam, Middlesbrough, Teesside TS 5 7 EH, Great Britain.

38. H. Maihak AG, D 2 Hamburg 39, Semperstr. 38, Germany.

39. Marine Electronics, Seattle, Wash., USA.

40. Merrick Scale Mfg. Co., 180 Autumn St., Passiac, N. J. 07055, USA.

41. Metritape Controls Inc., 77 Commonwealth Ave., West Conrad, Mass. 01781, USA.

42. MIAG- Mühlenbau- und Industrie GmbH., D 33 Braunschweig, Postfach 69, Germany.

43. Johannes Möller, D 2 Hamburg 50, Schillerstr. 53, Germany.

44. Netter Vibrationstechnik, D 62 Wiesbaden, Hasengartenstr. 40, Germany.

45. Nilos GmbH, D 4 Düsseldorf, Achenbachstr. 26, Germany.

46. Orthos Engineering Ltd., 62 Coventry Road, Market Harborough, Leicestershire, Great Britain.

47. Pneu-Bin Corp., 1970 E. 73rd. St., Chicago, Ill., USA.

48. Polysius AG, D 4723 Neubeckum, Postfach 340, Germany.

49. Redler Conveyors Ltd., Stroud, Glos. GL 5 3 EY, Great Britain.

50. Schaffer Poidometer Co., 2828 Smallman St., Pittsburgh 22, Pa. 15222, USA.

51. Schmelzbasaltwerk Kalenborn - Dr.-Ing. Mauritz KG, D 5461 Kalenborn, Germany.

52. Schwäbische Hüttenwerke GmbH, D 7083 Wasseralfingen, Postfach 25, Germany.

53. Siemens-Schuckertwerke AG, D 8520 Erlangen 2, Postfach 325, Germany.

54. Simon Handling Engineers Ltd., Stockport SK 3 ORX, Great Britain.

55. Stephens-Adamson Mfg. Co., Ridgeway Ave., Aurora, Ill. 60507, USA.

56. Richard Sutcliffe Engineering Systems Ltd., Harbury, Wakefield, Yorkshire, Great Britain.

57. The 30 - 98 Co. Ltd., Gatwick House, Povey Cross Road, Harley, Surrey, Great Britain.

58. Vibra Maschinenfabrik, Schultheis & Co. KG, D 605 Offenbach, Mühlheimer Str. 243, Germany.

59. Gewerkschaft Eisenhütte Westfalia, D 4628 Altlünen, Postfach 284-286, Germany.

60. Westfalia Dinnendahl Gröppel AG, D 463 Bochum, Postfach 2729, Germany.

61. Weinheimer Gummiwarenfabrik Weisbrod & Seifert GmbH, D 694 Weinheim, Postfach 169, Germany.

Series on Rock and Soil Mechanics
Vol. 1 (1971) No. 2

W. E. Dreyer
THE STRENGTH PROPERTIES OF ROCKS x)
=====================================

Contents

<u>Mechanics of Minerals</u>

Plasticity and strength of rock forming minerals, mechanisms of mineral deformation, strain hardening, elasticity of minerals.

<u>Mechanics of Rocks</u>

Cube compressive strength as a function of composition of minerals, grain sizes, fabric of grain boundaries and homogeneity of grains. Introduction of a "Mineral and Fabric Evaluation Degree", theoretical foundations for obtaining strength parameters, uni-, bi- and triaxial tests. Production and measurement of preferred orientations of minerals in rocks, determination and description of the textures.

<u>Mechanics of Rock Masses</u>

Compression mass-strength, model tests, measurement of absolute rock pressure, hydraulic recompression, load bearing capacity, convergence and age of bore-holes in elasto-plastic rock masses.

Price: US $ 25.00, Approx. 250 pp.

Available: April/May 1971

x) Note: This title was substituted for the planned book "The Science of Rock Mechanics" by the same author.

Series on Rock and Soil Mechanics
Vol. 1 (1971) No. 4

O. P. Natau

STATE-OF-STRESS MEASUREMENTS IN ROCK MASSES

Contents

The Fundamentals of Stress Measurements

Introduction, analysis of stress and strain, theoretical and practical considerations.

Instruments for Stress Measurements

Mechanical instruments, electrical instruments, hydraulic instruments, optical instruments, auxiliary equipment.

Methods of Measuring

Stress relief method, compensation method, hard inclusion method, special methods.

The Problem of Calibration

Calibration of instruments, calibration of methods.

Price: US $ 25.00, Approx. 250 pp.

Available: July 1971

Series on Rock and Soil Mechanics

Vol. 1 (1971) No. 5

Ch. Herbst

SLOPE STABILITY IN SOIL AND LOOSE ROCK - STATE OF THE ART AND FUTURE TRENDS -

Contents

The Mechanical Behavior of Soils and Loose Rock

Strength of soils, types of failure, volume changes, segregation effects.

Deformation Processes in Slopes of Soil and Loose Rock

Types of mass movements in slopes, influence of soil and loose rock characteristics, landslides, earthflows, mudflows, mass movements in naturally and artificially deposited soils.

The Stability Analysis of Slopes

Stability Analysis by computation, possibilities for calculation of strains and stresses, the limit equilibrium method, model tests.

Price: US $ 25.00, Approx. 250 pp., 400 references

Available: October 1971

Series on Tectonics in Teaching and other Projects

Edited by Professor Dr. Andreas Pilger
Director of the Geological Department
Technical University Clausthal, Clausthal-Zellerfeld, Germany

Clausthaler Tektonische Hefte

No. 1, 4. Ed. 1965; Adler, R., Fenchel, W., Hannak, W., Pilger, A.: Einige Grundlagen der Tektonik I. 64 pp., 39 Fig., 2 Tab., DM 6,20.

No. 2, 3. Ed. 1965; Adler, R., Fenchel, W., Pilger, A.: Statistische Methoden in der Tektonik I. - Die gebräuchlichsten Darstellungsarten ohne Verwendung der Lagenkugelprojektion. 97 pp., 51 Fig., 9 Tab., DM 9,80.

No. 3, 3. Ed. 1967; Adler, R., Fenchel, W., Martini, H.-J., Pilger, A.: Einige Grundlagen der Tektonik II. Die tektonischen Trennflächen. 94 pp., 67 Fig., 1 Tab., DM 7,60.

No. 4, 3. Ed. 1969; Adler, R., Fenchel, W., Pilger, A.: Statistische Methoden in der Tektonik II. Das Schmidtsche Netz und seine Anwendung im Bereich des makroskoptischen Gefüges. 111 pp., 79 Fig., DM 10,30.

No. 5, 1964; Karl, F.: Anwendung der Gefügekunde in der Petrotektonik, Teil I Grundbegriffe. 142 pp., 73 Fig., DM 16,20.

No. 6, 1967; Kronberg, P.: Photogeologie. Eine Einführung in die geologische Luftbildauswertung. 235 pp., 126 Fig., DM 23,-.

No. 7, 1968; Bottke, H.: Schürfbohren. Teil I: Die Durchführung von Kernbohrungen. 140 pp., 46 Fig., 60 Tab., DM 21,50.

No. 8, 1968; Adler, R., Krückeberg, F., Pfisterer, W., Pilger, A., Schmidt, M. W.: Elektronische Datenverarbeitung in der Tektonik. 157 pp., 23 Tab., 39 Fig., DM 22,50.

No. 9, 1969; Fuchs, W., Günther, R., Krausse, H.-F., Meyer, W., Mohr, K., Pertold, Z., Pilger, A.: Lineamenttektonik und Magmatismus im Westharz. 260 pp., 6 Tab., 32 Fig., DM 28,-.

No. 10, 1970; Computer-Einsatz in der Geologie. 400 pp., 126 Fig., 13 Tab., DM 54,-.

Clausthaler Geologische Abhandlungen

No. 1, 1965; Adler, R., Krausse, H.-F., Lautsch, H., Pilger, A.: Beiträge zur Tektonik des nördlichen Ruhrkarbons. 167 pp., 62 Fig., 1 Tab., DM 10,-.

No. 2, 1965; Krausse, H.-F.: Gefügeachsen im westlichen Teil des Vestischen Hauptsattels (nördliches Ruhrgebiet) und ihre Beziehung zu den übrigen tektonischen Elementen. 114 pp., 191 Fig., 7 Tab., DM 10,-.

No. 3, 1970; Wellmer, F.-W.: Verwendung piezoelektrischer Messungen an quarzhaltigen Gesteinen in der Gefügekunde. 190 pp., 60 Fig., 63 Tab., DM 15,-.

Published by
Verlag Ellen Pilger, Clausthal-Zellerfeld, Germany

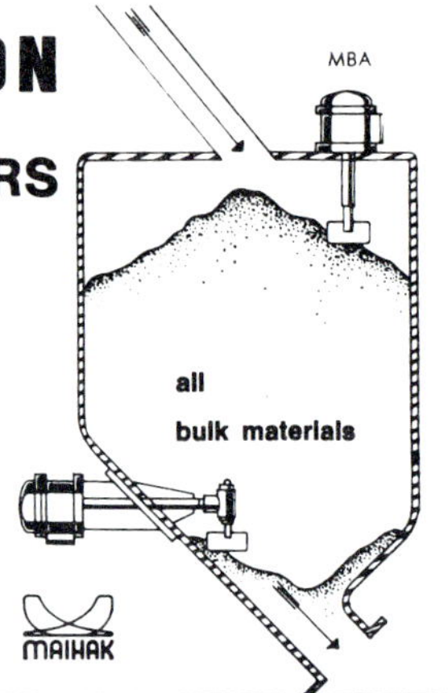

MBA
all
bulk materials

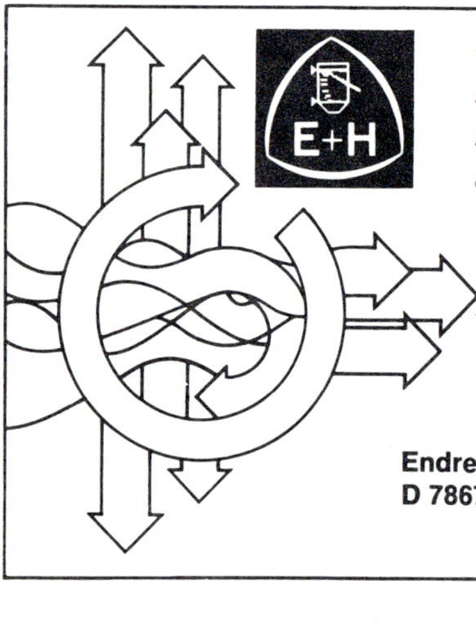